A CANAL BIBLIOGRAPHY WITH A PRIMARY EMPHASIS ON THE UNITED STATES AND CANADA

Compiled by

Albright G. Zimmerman, Ph.D.

Published by the

Canal History and Technology Press

Hugh Moore Historical Park and Museums

Easton, Pennsylvania

and

Pennsylvania Canal Society

1991

copyright 1991, by
Canal History and Technology Press

ISBN 0-930973-11-9

Published by
Canal History and Technology Press
Hugh Moore Historical Park and Museums, Inc.
P.O. Box 877, Easton, Pa. 18044-0877

in cooperation with the

The Pennsylvania Canal Society
P.O. Box 877, Easton, Pa. 18044-0877

All rights reserved
First Edition

Printed by
Garlits Printing
30 N. Pennsylvania Avenue
Morrisville, Pennsylvania
United States of America

A CANAL BIBLIOGRAPHY
With primary emphasis on the United States and Canada

Table of Contents
- Table of Contents.................................i
- Preface...iii
- Abbreviations.....................................1
- Bibligraphical Aids...............................1
- Periodicals and Newspapers (with continuing canal items..................................3
- Current and modern serials or Newsletters with canal information....................3
- General works....................................4
- General works- American..........................9
- Biographies.....................................25
- Pennsylvania....................................29
- New Jersey......................................67
  - Delaware & Raritan Canal.................67
  - Morris Canal.............................69
  - Other....................................70
- New York..73
- Ohio..86
- Illinois..91
- Indiana...95
- Kentucky..97
- Michigan..98
- Wisconsin.......................................99
- Massachusetts..................................100
- Connecticut....................................103
- Rhode Island...................................104
- New Hampshire..................................104
- Vermont..105
- Maine..105
- Maryland.......................................106
  - Chesapeake and Delaware Canal..........106
  - Chesapeake and Ohio Canal..............107
  - Other..................................110
- Delaware.......................................110
- Washington, D.C................................111
- Virginia.......................................111
- West Virginia..................................114
- South Carolina.................................114
- Georgia..115
- Alabama..115
- Tennessee......................................116
- Tennessee-Tombigbee Waterway...................116
- Texas..117
- North Carolina.................................117
- Louisiana......................................118
- Florida..119

--- i ---

Table of Contents (Cont.)

Gulf Intercoastal Waterway..................120
Atlantic Intercoastal Waterway..............121
California..................................122
Washington (state) and Oregon...............123
Mississippi-Ohio Canalized River System....124
TVA...129
Sault Ste. Marie............................130
St. Lawrence Seaway.........................132
Canada......................................134
Panama Canal................................142
Nicaragua...................................152
Techuantepec Ship Railway...................154
Suez Canal..................................155
British Canals (selected)...................158
    British Bibliographical and General...158
        England...............................160
            Scotland..........................161
            Ireland...........................162
European- unclassified......................163
France......................................165
Spain.......................................167
Belgium.....................................168
Netherlands.................................168
Germany.....................................169
Italy.......................................170
Sweden......................................170
Greece......................................171
Roumania....................................171
Finland.....................................171
Poland......................................171
Russia......................................171
Philippines.................................173
India.......................................173
Japan.......................................173
Thailand....................................173
Madagascar..................................174
Australia...................................174
Vietnam.....................................174
Columbia....................................174
Mexico......................................174
South America (general).....................174
China.......................................175
Sudan.......................................176
Ancient Canals..............................177
Medieval and Renaissance Canals.............179
Appendix A..................................180
Appendix B..................................186

Preface

As of this moment, this is a bibliography of canal history and can be best described as "in progress". It was initiated by necessity as I worked on various projects relating to Pennsylvania canal history and has continued to grow. As I collected bits and pieces here and there, it became apparent that adequate or even useful canal bibliographies did not exist. This, I discovered, was particularly the case in reference to secondary works. Encouraged by Lance Metz, historian for the Hugh Moore Historical Park and Museums and the primary motivator for their canal symposiums, it was decided that I should expand my bibliography. The expansion extended my collecting to North America and ultimately, somewhat selectively to the entire world. Lance and I have decided that the bibliography, which has reached something in excess of 20,000 entries, has reached a stage where it would be useful to both scholars and canal buffs and already it should begin to fill the void that now exists.

With the exception of dissertations and a handful of other carefully noted unpublished but available works, every other work listed; book, pamphlet and periodical article, is secondary. The majority are twentieth century although there are a sizeable number of nineteenth century items included. Where they exist, modern reprints are noted. Most of the material, although not necessarily all, can be located in major libraries and is available through inter-library loan.

It has been necessary to be selective and even arbitrary in selecting the entries. Many articles that appeared to be superficial to the compiler, for example many of those in fishing and outdoors magazines and which in many cases were difficult to obtain were rejected out of hand. There were however, among the rejects, many brief articles and notices that were dismissed simply because of their brevity or because they were "newsy" items about field trips or post canal-history concerns. In some cases, brief items and notices are included, because in the judgement of the compiler, they make a significant contribution, either because of their unique historical importance or because in many cases they were the only items discovered dealing with specific foreign canals or constituting the only discoverable notices of forgotten American canals.

Despite its 20,000 plus entries, this bibliography is by no means definitive. For example, there are an incredible number of published source documents available and no effort has been made to include any such materials. However, bibliographies are listed and in Adelaide Hasse's bibliography for Pennsylvania one can find more than a hundred pages cataloguing published governmental sources dealing with rivers and canals that refer to hundreds of thousands of pages of reports and minutes falling primarily between 1825 and 1860.

In the process of compiling these listings, I have made runs through complete files of *Technology and Culture* (with its yearly bibliographies), the English *Transactions of the Newcomen Society, Inland Seas, The Journal of the Franklin Institute, The Pennsylvania Magazine of History and Biography, Pennsylvania History, The Bulletin of the Pennsylvania Department of Internal Affairs, Niles Register, Hazard's Register(s), Railroad History* [and its predecessors], *New Jersey History* [and its predecessors], the English *The Journal of Transport History* and the English *Transport History* as well as the publications of several of the Pennsylvania county historical societies. In addition, wherever available, index volumes for various historical periodicals such as those of Michigan, Indiana and New York, and were perused along with those of the *Mississippi Valley Historical Review* [now the *Journal of American History*]. A great many but certainly by no means all of the items were physically examined by this compiler. Among the richest sources were the complete file of the Pennsylvania Canal Society's *Canal Currents* and the file of the American Canal Society's *American Canals*, both of which were examined at the Support Center of Easton's Hugh Moore Historical Park and Museums.

As it may be difficult to find files of these latter publications, for the information of anyone interested, most of the back issues of *Canal Currents* are still available for purchase from the Pennsylvania Canal Society through Hugh Moore Park and if the interest is there, a full run can be obtained with several issues made available through xeroxing. Some similar arrangements should be possible for *American Canals* through the American Canal and Transportation Center in York, Pennsylvania. These are among the richest general sources available although there are in existence a few definitive volumes for some of the canals and state systems. For example, one might note Petrillo's history of Pennsylvania's North Branch Canal, Grey's study of the Chesapeake and Delaware Canal and Sanderlin's history of the C & O canal. A few state systems have also been adequately covered in works such as Scheiber's study of the Ohio Canals, Fatout's Indiana canal history and the works of Leggett and Passford on the Canadian Canals. Particularly worthy of mention is Ronald Shaw's recent comprehensive history of American Canal between 1820 and 1860. Too many canals and state efforts are still lacking authoritative histories and in too many cases any real history. It is my hope that this bibliography may serve as a tool to help scholars fill the void.

I make no apologies for including many of the accounts of "antiquarians" along with those of the most sophisticated "cliomatricians" [the somewhat rarified proponents of the "new history" with its formulas, quantifications and computer applications]. Each, along with the numerous more main-stream historians, has something to contribute to the canal historian. In this light, I have included the range from "folklore" to "sophistication".

New Yorkers and Ohioans will probably consider me chauvinistic as I have included the Delaware and Hudson Canal and the Pennsylvania and Ohio Canal respectively in the Pennsylvania listing. Canal buffs from other states will probably look askance at the listing of the Chesapeake and Ohio Canal and the Chesapeake and Delaware Canals under Maryland, but that's where they fell as I developed my bibliography and there at least for the present, they will remain.

Elsewhere, I have included everything dealing with interoceanic isthmian canals under the Panama Canal unless it specifically referred to Nicaragua or the Tehuantepec Ship Railway.

I have really just started an examination of such publications as *Scientific American*, *Engineering News*, the *Engineering News-Record* and the *Transactions of the American Society of Civil Engineers*. These volumes, only sampled to date, are a gold mine of detailed engineering and historical information. Unfortunately many pertinent items are mere notices and had to be rejected out of hand because of their brevity, but there are additional numerous rewarding entries of greater detail, usually accompanied by engineering sketches, worthy of inclusion. Hopefully another addition will reflect the riches of these sources.

Notice should also be taken of some of the fabulous electronic tools that are now available to the researcher and the bibliographer. I refer particularly to the OCLC on line that catalogues and locates more than 20,000,000 titles and their listings are increasing at a monumental rate. There are competitors that probably are equally useful and equally worthy.

We have left fairly wide margins and occasional portions of pages blank as an invitation to users to make notes and additions as they use the bibliography.

As already noted, this is a "progress report" and in this light I solicit your responses, both positive and negative, to my selections, as well as your suggestions for any additions that I should be aware of. It is impossible in a listing with 20,000 plus entries, to have examined every item and annotation would have been possible only if I had severely limited my selections.

I hope there will be an opportunity for a revised edition in the future, possibly with cross referencing and limited annotation enriched, I hope by contributions from many of you, and I further hope that this compilation will aid you in your researches and stimulate you and your students or disciples to new historical efforts.

Rider College   Albright G. "Zip" Zimmerman

home- 1361 River Road   or   c/o Hugh More Historical Park and
      Yardley, PA 19067   Canal Museums
                                                     P.O. Box 877
                                                     Easton, PA 18042-0877

A CANAL BIBLIOGRAPHY
With primary emphasis on the United States and Canada

Abbreviations used:

AC American Canals, Bulletin of the American Canal Society
CC Canal Currents: Bulletin of the Pennsylvania Canal Society
HR Hazard's Register of Pennsylvania
PDIA Pennsylvania Department of Internal Affairs Bulletin
PH Pennsylvania History
PMHB Pennsylvania Magazine of History and Biography
CCHT Canal History and Technology Proceedings
WPHM Western Pennsylvania Historical Magazine

BIBLIOGRAPHICAL AIDS:

Dictionary Catalogue of The Research Libraries of the New York Public Library, 1911-1971 (New York Public Library, printed and distributed by G.K. Hall & Co.), vol. No. 113 includes "canals".

Engineering Societies Library [New York], *Classed [sic.] Subject Catalogue,* 13 volumes (Boston: G.K. Hall, 1963), particularly vol. 1, 465-467, vol. 8, 530-610, vol. 12, 449-451; 10 supplements, 1964- 1973.

Ferguson, Eugene S., *Bibliography of the History of Technology,* Society for the History of Technology Series, No. 5 (Cambridge, Mass: MIT Press, 1968).

Historic Writings on Hydraulics: A Catalogue of the History of Hydraulics Collection in the University of Iowa Libraries (Iowa City: Friends of the Univ. of Iowa Libraries, 1984).

Hoornstra, Jean and Trudy Heath (eds.), *American Periodicals 1741-1900, An Index to the Microfilm Collections* (Ann Arbor, Mich: University Microfilms International, 1979).

Larson, Henrietta M. (comp.), *Guide to Business History* (Cambridge, Mass: Harvard Univ. Press, 1948)

The National Union Catalogue Pre-1956 Imprints, 753 vols. (London: Mansell Information/ N.Y: American Library Association, 1976) [there are various other editions and various publishers plus chronological supplement series].

BIBLIOGRAPHICAL AIDS: (cont.)

Peddie, R.A., *Subject Index of Books Published up to and Including 1880*, Second series (London: Grafton & Co., 1935) ["Canals and Inland Navigation, 125-126], other editions available.

Poole, William Frederick, *Poole's Index to Periodical Literature*, 6 vols, volume I alphabetical (1802-1881), vols II through VI, Chronological, 1882-1906 (1882, 1938, 1958, Gloucester, Mass: Peter Smith, 1963

Stapleton, Darwin H. (comp.), *Accounts of European Science, Technology, & Medicine Written by American Travelers Abroad, 1735-1860, in the Collections of the American Philosophical Society* (Philadelphia: American Philosophical Library, 1985), contains references to observers of canals and related technology.
——————————————, *The History of Civil Engineering Since 1600: An Annotated Bibliography* (New York and London: Garland Publishing Inc., 1986).

Technology and Culture, yearly bibliographies, various compilers, Vol III (1962) to date.

PERIODICALS and NEWSPAPERS [with continuing canal items]

The American Almanac and Repository of Useful Knowledge, annual volumes, 1830-1861, Boston and New York
American Railroad Journal, 1832-???
Engineering News and *Engineering News-Record*
Hazard, Samuel (ed.), *The Register of Pennsylvania devoted to the Presentation of Every Kind of Useful Information respecting the State*, 16 volumes, 1828-1836.
Hazard's United States Commercial and Statistical Register, 6 vols., 1840-1842.
Hunt's *Merchant Magazine and Commerical Review*, 1839-1870.
Journal of the Franklin Institute, 1826-date.
Niles Weekly Register, (Baltimore, 1811-1849).
National Waterways, from about 1928-1932,
Scientific American, from   to about 1920.

Current and/or modern serials or Newsletters with canal information
American Canals, Bulletin of the American Canal Society, 1972-present.
Bottoming Out, Canal Society of New York State, Nos. 1- 18/9, 1956-1965
Canal & Riverboat, monthly, British), published by A.E. Morgan Pub. Ltd., Stanley House, 9 West Street Epsom, Surry KT18 7RL
Canal Currents, Canal Society of Pennsylvania, 1968-present.
The Canal Packet: Newsletter of the Canal Museum Associates, 1977-present [New York].
Canals Canada: Newsletter of Canadian Canal Society Jan 1983-????.
Indiana Waterways, Canal Society of Indiana.
Inland Seas, Quarterly Bulletin of the Great Lakes Historical Society.
Journal of Transport History [British] 196 -1970; new series 1971-1979; 3d series 1980 to date. .
The Locktender, Newsletter of Hugh Moore Park and Canal Museums.
Now and Then [Muncey, Pennsylvania]
Proceedings of the Center for Canal History and Technology [annual]
The Tiller, Virginia Canals & Navigation Society, 1980 to present.
Transport History, [British] 1968 to date.
The Towpath Post: Journal of the Canal Society of New Jersey, 1969-????
Towpaths: the Quarterly of the Canal Society of Ohio, 1961-????
Waterways World, [monthly British publication for canal users]

GENERAL WORKS

Bathe, Greville, *An Engineer's Miscellany* (Philadelphia: Patterson & White Co., 1938), Chapter VII, "The Antiquity of the Inclined Plane on Canals."

Baxter, R.R., *The Law of International Waterways With Particular Regard to Interocean Canals* (Cambridge, Mass: Harvard Univ. Press, 1964).

Boulton, W.H., *The Pageant of Transport Through the Ages* (New York and London: Benjamin Blom, reissued 1969), Chapters VIII and IX.

Buchanan, R.A., "The Diaspora of British Engineering," *Technology and Culture*, XXVII (1986), 501-524.

"Canal Locks and Inclines," *AC*, No. 41 (May 1982), 5 [from *Engineering and Railroad Journal*, I, June 8, 1867].

Chaloner, W. H., "John Phillips [c.1740-1820]: Surveyor and Writer on Canals," *Transport History*, V (1972), 168-172 [England and Russia].

Chiles, Webb, "The Two Great Canals," (Panama and Suez), *Sail*, XIX (Feb 1988), 63-67.

Davidson, Frank P., L.J. Giacoletto and Robert Salkeld (eds.), *Macro-Engineering and the Infrastructure of Tomorrow* (Boulder Col: Westview Press for the American Association for the Advancement of Science, 1967). [scattered items on canals of the past, present and future].

Davidson, Frank P., *Macro: A Clear Vision of How Science and Technology Will Shape Our Future* (New York: William Morrow and Co., 1983) [scattered items on canals, past present and future].

Daumas, Maurice (ed.), *A History of Technology and Invention: Progress Through the Ages*, translated by Eileen B. Hennessy. Vol I "The Origins of Technological Civilization," 131, 149. 226, 305; Vol. 2, "The First Stages of Mechanization," 127-30, 421-429 (1962, 1964, English translations, New York: Crown Publishers, Inc., 1969).

Evans, Francis T, "Railways and Canals: Technical Choices in 19th Century Britain," *Technology and Culture*, XXII (1981), 1-34.

Fairlie, J.A., "The Economic Effects of Ship Canals," *Annals of the American Academy of Political and Social Science*, XI (1898), 54-78.

GENERAL WORKS–(cont.)

Flaxman, Edward, *Great Feats of Modern Engineering* (Freeport, N.Y: Books for Libraries. Inc., 1938, reprint 1967), pps 80-119.

Fulton, Robert, *A Treatise on the Improvement of Canal Navigation* (London: I & J Taylor, 1796).

Funk, Jeffrey L., "A Methodology for Estimating the Benefits from New Lock Construction," *Transportation Quarterly*, XXXVII(4) (Oct 1983), 597-621.

"The Great Canals of the World," *National Geographic*, XVI (1905), 475-479; based on O. P. Austin, "The Great Canals of the World," Bureau of Statistics of the Department of Commerce and Labor (Washington, D.C: Gov't Printing Office, 1902).

Gresswell, R. Kay and Anthony Huxley, *Standard Encyclopedia of the World's Rivers and Lakes* (New York: G. P. Puthnam's Sons, 1965), useful references to canalized streams and connecting artificial waterways.

Hadfield, Charles, "Evolution of the Canal Inclined Plane," *AC*, No. 56 (August 1986),4-7, reprint from *RACHS Journal*, September, 1979.
———————————, *World Canals* (Newton Abbot: David & Charles, Inc., 1986; also New York & London: Facts on File Publishers, 1986)

Harris, Robert, *Canals and Their Architecture* (New York: Frederick A. Praeger, 1969), also (London: H. Evelyn, 1969).

Haupt, Lewis M., "Commerce and Deepways," *Journal of the Franklin Institute*, CXLI (1896), 80-97; 171-182 [illustrated].
———————————, "Ship Canals," *Journal of the Franklin Institute*, CXXXIV (1892), 339-353.

Hepburn, A. Barton, *Artificial Waterways of the World* (New York: Macmillan, 1909, 1914 and 1918).

Hess, W.N., "New Horizons in Resource Development: The Role of Nuclear Explosions," *Geographical Review* LII (1962), 1-24 [canals, 9-16].

Hoskins, Kenneth T., "The Navigational Lock," *Transport History*, XII (1981), 11-15.

Jacobs, David and Anthony E. Neville, *Bridges, Canals and Tunnels* (New York: American Heritage in association with Smithsonian Institution, 1968).

GENERAL WORKS- (cont.)

Jeans, J. Stephen, *Waterways and Water Transport in Different Countries* (London and New York: E & F. N. Spoor, 1890).

Jensen, Martin, *Civil Engineering Around 1700: With special reference to Equipment and Methods* [published in Danish and English on the occasion of the 50th anniversary of Monberg & Thorson] (Copenhagen: Danish Technical Press, 1969)

Kirby, Richard S., Sidney Withington, Arthur B. Darling and Frederick G. Kilgour, *Engineering in History* (New York, Toronto, London: McGraw-Hill Book Company, 1956), 33, 113-114, 140-145, 207-220, 444-456.

Leighton, Albert C., "The Mule as a Cultural Invention," *Technology and Science*, VIII (1967), 45-52.

MacElwee, Roy S., "Ship Canals and Canal Locks," *National Waterways*, X (April 1931), 10-13, 32.

McNown, John (ed.), "Canal Hydraulics-a la Franklin," *AC*, No. 35 (November 1980), 6.

Mance, Osborne, *International River and Canal Transport* (London, New York and Toronto: Oxford Univ. Press, 1945)

Mann, Theod. Aug., "XXXVII. A Treatise on Rivers and Canals," [read June 24, 1779], *Philosophical Transactions*, LXIX (1779), 555-607; [facsimile reprint, 1765-1780, I to LXX, New York: Johnson Reprint Corporation/Kraus Reprint Corporation, 1963].

Morris, Ellwood, "On the Mensuration of Excavation and Embankment, upon Canals, Roads, and Rail Roads," *Journal of the Franklin Institute*, XXIX (1840), 21-35.

Newcomer, Henry C., "Improvement in Rivers for Navigation, " *Engineering News*, LXVII (May 30, 1912), 1033-1036.

O'Brien, W., "Canals and Canal Conveyance," *Journal of the Franklin Institute*, LXVII (1859), 17-26, 73-80.

Pannell, J.P.M., *An Illustrated History of Civil Engineering* (New York: Ungar, 1964), Ch II, "Rivers and Canals, pps 45-90.

Payne, Robert, *The Canal Builders* (New York: Macmillan Co., 1959).

Pelts, Geoff, "Mighty Rivers of the World: How Man has Depended On, Tamed and Abused Them," *Geographical Magazine*, LIX (1987), 434-443.

GENERAL WORKS (cont.)

Pilkington, Roger, "Canals: Inland Waterways Outside Britain," in Singer, Charles, E.J. Holmyald, A.R. Hall and Trevor T. Williams (eds.), *History of Technology*, 5 vols (Oxford Univ Press, 1955-1958), IV, Chapter 18, part 1..

Provis, W.A., "On the Locks Commonly Used for River and Canal Navigation," *Journal of the Franklin Institute*, XXIII (1837), 218-223.

Rubin, Norman N., "Canal Boats," *Nautical Research Journal*, XV(2) (1968), 71-79.

Shank, William H., Philip Ogden, Terry Woods and William Dzombak, "Saving Water at the Locks," *AC* No. 71 (November 1989), 7, 10.

Shempton, A.W., "Canals and River Navigation before 1750," in Singer, Charles, et al (eds.), *History of Technology*, 5 vols. (Oxford University Press, 1955-1958), III, Ch. 15.

"Ship Canals in 1889," *Engineering News and American Railway Journal*, XXII (July-December 1889), 100-103,126-128.

"Sketch of the Progress of Inventions Connected with Navigable Canals," *Journal of the Franklin Institute*, VI (1828), 188-193; 228-239, 295-301.

Smiles, Samuel, *Lives of the Engineers with an Account of their Principal Works comprising also A History of Inland Communication in Britain*, 3 vols., (London: John Murray, 1862, reprint. Augustus M. Kelley, 1968).

Smith, Norman, *A History of Dams* (Secaucus, N.J: The Citadel Press, 1971, 1972), scattered items.

Springer, J.F., "Some Great Dredges," *Scientific American*, CXXIX (1923), 394-395.

(T) [Treadwell, Daniel], "Sketches of the Progress of Inventions, connected with Navigable Canals. Compiled from various sources," *The Boston Journal of Philosophy and the Arts*, I (1823-4), 479-91, 530-87.

Tew, David, *Canal Inclines & Lifts* (London: Alan Sutton Publications, Ltd., 1984)
----------, "Canal Lifts and Inclines With Particular Reference to Those in the British Isles," *Newcomen Society Transactions* [English], XXVII (1951-1952) and (1952-1953), 35-58.

GENERAL WORKS (CONT.)

Trout, William E., "The Martian 'Canals'," *AC*, No. 28 (February 1979), 6-7.

Turnbull, Gerard, "Canals, Coal and Regional Growth During the Industrial Revolution," *The Economic History Review* [British], 2d series, XL (1987), 537-560.

----------------, "From Thames to Titicaca: An Appreciation of Charles Hadfield," *Journal of Transport History* VIII(1) (1987), 99-102.

Turner, Roland and Steven L. Goulden (eds.), *Great Engineers and Pioneers in Technology*, vol I, (New York: St. Martins, 1982).

Vernon-Harcourt, L. F., *Rivers and Canals*, 2 vols. (Oxford, Clarendon Press, 1882, rev. ed., 1896).

JUVENILE

Bean, Keith F., *Famous Waterways of the World*, rev. ed. (London: F. Muller/New Rochelle: N.Y., Sport/Shelf, [c1956], 1963).

Boyer, Edward, *River and Canal* (New York: Holiday House, 1986).

Morrison, Frank, *Golden Ditches: a Story of Canals, Rivers and Waterways* (Minneapolis, Minn: T.S. Denison, 1970).

Russell, Solveig P., *The Big Ditch Waterways: The Story of Canals* (New York: Parents Magazine Press, c. 1977).

Sandak, Cass R., *Canals* (New York: F. Watts, 1983).

Scarry, Huck, *Life on a Barge* [European]: *A Sketchbook* (Englewood Cliffs, N.J: Prentice Hall, 1982).

GENERAL WORKS- AMERICAN

Bibliographical Aids

Feltner, Charles E. and Jeri Baron Feltner, *Great Lakes Maritime History: Bibliography and Sources of Information* (Dearborn, Mich: Seajay Publications, 1982).

Historic American Engineering Record, *HAER Checklist, 1969-1985* (National Park Service, P.O. Box 37127, Washington, D.C., 20013, 1985)

Hoy, Suellen M. and Michael C. Robinson (compilers and eds.), *Public Works History in the United States: A Guide to the Literature* (Nashville, Tenn: American Association for State and Local History, 1982).

Madden, Emily A., *Index of Canal References in Niles' Weekly Register volume 1-76* (Livonia, N.Y: author, 1982).

Nelson, Daniel (comp.) *A Checklist of Writings on the Economic History of the Greater Philadelphia-Wilmington Region* (Greenville, Del: Eleutherian Mills Historical Library, 1968).

Poore, Ben: Perley (comp.), *A Descriptive Catalogue of The Government Publications of the United States, September 5, 1774-March 4, 1881* (48th Congress, 2d Series, Senate Misc. Doc. No. 67 (Washington: Govt. Printing Office, 1885, reprint, Ann Arbor, Mich: J. W. Edwards Publisher, Inc., 1953) indexed, numerous items under Canals.

Rink, Evald, *Technical Americana: A Checklist of Technical Publications before 1831* (Millwood, N.Y: Kraus International Publications, 1981).

Shaw, Ralph .R., *Engineering Books Available in America Prior to 1830* [with introductory notes on early engineering education in America], *Bulletin of the New York Public Library*, XXXVII (1933), 38-61 and monthly to 539-561.

Shaw, Ronald E., "Canals in the Early Republic: A Review of Recent Literature," *Journal of the Early Republic*, IV (1984), 117-142.

Thompson, Thomas R., *Check List of Publications on American Railroads before 1841* (New York: New York Public Library, 1942).

GENERAL WORKS—AMERICAN (Cont.)

Other Works

 Armoyd, George, *A Connected View of the Whole Internal Navigation of the United States* (1830, reprint, Burt Franklin, 1971).

 Armstrong, Ellis, Michael C. Robinson and Suellen M. Hoy, (eds.), *History of the Public Works in the United States, 1776-1976* (Chicago: Public Works Association, 1976).

 Ault, Phillip H., *Whistle Round the Bend: Travel on America's Waterways* (New York: Dodd, Mead, 1982)

 Baer, Christopher T, et al, *Canals and Railroads of the Mid-Atlantic States, 1800-1860* (Wilmington, Del: Regional Economic History Research Center Eleutherian Mills-Hagley Foundation, 1981).

 Baldwin, Laommi, *Thoughts on the Study of Political Economy as Connected with the Population, Industry and Paper Currency of the United States* (1809, reprinted, Augustus M. Kelley, 1968)

 Barnard, Charles, "Inland Navigation of the United States," *Century*, XXXVIII (1889), 353-372.

 Brazer, Marjorie Cahn, *Cruising Guide to the Great Lakes and Their Connecting Waterways* (Chicago, Ill: Contemporary Books, 1985)

 Brown, Alexander C., "Reversible Head Locks," *AC*, No. 34 (August 1980), 3.

 Cairo, Robert F., "Materials Towards an History of Early Hydraulic Engineering Machinery Employed in the Development of Rivers, Harbors and Canals," "Part 1, " *Nautical Researh Journal*, XXVIII (June 1982), 10-62, 100; "Part 7," XIX (Dec 1983), 185-194; "Part 8," XXX (June 1984), 77-??.

 Calhoun, Daniel H., *The American Civil Engineer: Origins and Conflict* (Cambridge, Mass: MIT Press, 1960).

 Callender, G. S., "The Early Transportation and Banking Enterprises of the States in Relation to the Growth of Corporations," *Quarterly Journal of Economics*, XVII (1903), 111-162, reprinted in Joseph T. Lambie and Richard V. Clemence (ed.), *Economic Change in America* (Harrisburg: Stackpoole, 1954), 522-559..

 Carter, Edward C., *Benjamin Henry Latrobe and Public Works: Professionalism, Private Interest, and Public Policy in the Age of Jefferson* (Washington D.C: Public Works Historical Society, 1976).

GENERAL WORKS—AMERICAN (cont)

Chevalier, Michael, *Histoire et description de voies des communication aux Etats Unis et des travaux qui en dependent*, 2 vols. (Paris: Librarie de Charles Gosselin et Ce, 1840-1841).

————————, *Lettres sur L'Amerique du Nord*, 2 vols. (Paris: Librairie de Charles Gosselin et Ce, 1837)

———————— *Society, Manners and Politics in the United States* trans. from 3d Paris Edition (Boston: Weeks, Jordan and Company, 1839)

Clark, Dennis, *Hibernia America: The Irish and Regional Cultures* (New York, Westport, Conn. and London: Greenwood Press, 1986), Chapter II, "Diggers".

Cochran, Thomas C. (general editor), *The New American State Papers: Transportation*, 7 vols, vols 3, 4, and 5 specifically on canals (Wilmington, Del: Scholarly Resources, 1972).

Cohn, Michael B., "The Early Economic Contributions of the Canals," *CC*, No. 64 (Autumn 1983), 15-16.

Cranmer, H. Jerome, "Canal Investment, 1815-1860," in William N. Parker (ed.), *Trends in the American Economy in the Nineteenth Century* (Princeton: Princeton Univ. Press, 1960).

Davis, Lance E. et al, *American Economic Growth: An Economic History of the United States* (New York et al: Harper & Row, 1972), "Age of Canal Expansion, 1815-1843," 476-485.

Davis, Joseph S., *Essays in the Earlier History of American Corporations*, 2 vols. (Cambridge, Mass., Harvard Univ. Press, 1917, reprint, Russell & Russell, 1965); "Corporations for Improving Inland Navigation," II, 109-185; Appendix B, 335-338.

Davis, W.W., "A Trip From Pennsylvania to Illinois in 1857," *Transactions of the Illinois State Historical Society*, Publication No. 9 (1904), 198-204, [description of canal packet].

Davison, Ann. *In the Wake of the Gemini* (Boston: Little Brown & Co., 1962), 6000 miles on America's inland waterways.

"The Deep-Water Route from Chicago to the Gulf," *National Geographic*, XVIII (Oct. 1907), 679-685.

Douglas, Paul H. and William K. Jones., "Sandstone, Canals, and the Smithsonian," *Smithsonian Journal of History*, III (Spring, 1986), 41-58.

Drago, Harry Sinclair, *Canal Days in America* (New York: Charles N. Potter, 1972).

GENERAL WORKS—AMERICAN (cont)

Dunbar, Seymour, *A History of Travel in America* (Indianopolis: Bobbs-Merrill, 1915; reprint, 1 volume, New York: Tudor Press, 1937; 4 volumes, New York: Greenwood Press, 1968).

Dzombak, William, "Canals and the Birth of Geology," *CC*, No. 55 (Summer 1981), 12-15.
————————, "Canals or Roads for America," *CC*, No. 65 (Winter 1984), 3-7.
————————, "Glacial Geology and Canal Building," *CC*, No. 62. (Spring 1983), 3-4, 11.

Ellis, Horace, "America's Great Canal Boom," *National Republic*, XIX (May 1931), 24-5, 44.

Finch, James Kip, *Early Columbia Engineers; an Appreciation* (New York, Columbia Univ. Press, 1929).
————————, "A Hundred Years of American Civil Engineering, 1852-1952," in American Society of Civil Engineers *Centennial Transactions*, vol. CT (1953).

Fishlow, Albert, Review of *Canals and American Economic Development* by Carter Goodrich and *Canal or Railroad?* by Julius Rubin, *Journal of Economic History*, XIII (1963), 129-131.

Fite, Emerson D., "The Canal and the Railroad from 1861-1865," *Yale Review*, XV (1906), 195-213.
————————, *Social and Industrial Conditions in the North During the Civil War* (1909, republished, New York: Frederick Ungar, 1963), particularly 45-54.

Ford, Alexander H., "Waterways of America," *Harpers*, CI (1900), 783-794.

Funk, Jeffrey L., "A Methodology for Estimating the Benefits from New Lock Construction," *Transportation Quarterly*, XXXVII (4) (Oct 1983), 597-621.

Gallatin, Albert, *Report of the Secretary of the Treasury on the Subject of Public Roads and Canals* (Washington, D. C: 1808; reprints, *American State Papers*, XX, Miscellaneous, I, 724-921; New York: Augustus M. Kelley, 1968).

Gates, Paul W., *History of Public Land Law Development* (Washington, D.C: Government Printing Office, 1968), Chapter XIV, "Land Grants for Railroads and Public Improvments."

GENERAL WORKS–AMERICAN–(Cont.)

Gephart, William F., *Transportation and Industrial Development in the Middle West* (Columbia University Studies in History, Economics and Public Law, XXXIV, No. 1, 1909), Ch. VII, "The Relation of Canals to the Industrial Development of the Middle West," 107-128..

Gerstner, Franz Anton: Ritter von., *Die Innern Communicationen der Vereinigten Staaten von Nordamerica*, aufgesetetzt, redigert und brag. von L. Klein, 2 vols. (Vienna: Forster's artisische Anstalt, 1842-1843).

Gerstner, Francis Anthony Chevalier de, "Letters from the United States of North America on Internal Improvements, Steam Navigation, Banking, &c," translated by L. Klein, *Journal of the Franklin Institute*, XXX (1840), 217-227; 289-301; 361-369; XXXI (1841), 73-82; 165-173.

Gerstner, Frantz Joseph, Ritter von, *Memoire sur les grandes routes, les chemins de fer et les canaux de navigation*: traduit de l'allemand, de F. M. de Gerstner. . .et predede d'une introduction, par M. P. S. Girard. . . (Paris: Bachelier, successeur de Mme Ve Courcier, libraire pour sciences. . ., 1827) [Girard's introduction describes Society for Promotion of Internal Improvements, etc].

Goodrich, Carter, "American Development Policy: the Case for Internal Improvment," *Journal of Economic History*, XVI (1956), 449-460.

———————— (ed.), *Canals and American Economic Development* by Carter Goodrich, Julius Rubin, H. Jerome Cranmer and Harvey D. Segal, (New York: Columbia Univ. Press, 1961).

————————, "The Gallatin Plan After One Hundred and Fifty Years," *Proceedings of the American Philosophical Society*, CII (1958), 436-441.

———————— (ed.), *The Government and the Economy, 1783-1861* (Indianapolis: Bobbs-Merrill, 1967).

————————, *Government Promotion of American Canals and Railroads, 1800-1890*, (New York: Columbia Univ. Press, 1960.

————————, "Internal Improvements Reconsidered," *Journal of Economic History*, XXX (1970), 289-311.

————————, "Local Planning of Internal Improvements," *Political Science Quarterly*, LXVI (1951), 411-445.

————————, "National Planning of Internal Improvements," *Political Science Quarterly*, LXIII (1948), 16-44.

————————, "Public Spirit and American Improvements," *Proceedings of the American Philosophical Society*, XCII (1948), 305-309.

————————, "The Revulsion Against Internal Improvements," *Journal of Economic History*, X (1950), 145-169.

GENERAL WORKS–AMERICAN–(CONT.)

Gray, Ralph D., "Transportation and Brandywine Industries," *Delaware History*, IX (1961), 303–325.

"Great Lakes Time Line [chronology]," *Inland Seas*, XL (1984), 175–197.

Hadfield, Charles and Alice Mary, *Afloat in America* (Newton Abbot, London and North Pomfret, Vt: David & Charles, 1979).

Haeger, John D., *The Investment Frontier: New York Businessmen and the Economic Development of the Old Northwest* (Albany: State University of New York, 1981)

Hall, James, *The West: Its Commerce and Navigation* (1848, reprint, New York: Burt Franklin, 1970).

Harlow, Alvin F., *Old Towpaths: The Story of the American Canal Era* (New York: D. Appleton & Co., 1926)
————————, *When Horses Pulled Boats: A Story of Early Canals* with an Inroduction by William H. Shank, reprinted and abridged (York, Pa: American Canal and Transportation Center, 1983).

Harrison, Joseph H., Jr., "*Sic et Non*: Thomas Jefferson and Internal Improvement," *Journal of the Early Republic*, VII (1987), 339.

Herndon, G. Melvin, "A Grandiose Scheme to Navigate and Harness Niagara Falls," *The New York Historical Society Quarterly*, LVIII (1974), 6–17.

Heydinger, Earl J., "Canals 'Hurting' Canals," *CC*, No. 55 (1981), 11.

Hill, Forest G., *Roads, Rails and Waterways: The Army Engineers and Early Transportation* (Norman, Okla: Univ of Oklahoma Press, 1957).

Hulbert, Archer B., "For the Trade of the Golden West," Part One, "The Erie and Its Rivals," *National Waterways*, VI (January 1929), 29–34, 54–56; Part Two, "The Conquest of the Alleghenies," VI (April 1929), 41–45.
————————, *Great American Canals*, volume 13 of Historic Highways of America Series (Cleveland, Ohio: Arthur H. Clark, 1904).
————————, *The Paths of Inland Commerce* (New Haven: Yale Univ. Press, 1920).

GENERAL WORKS–AMERICAN (Cont.)

Hull, William J. and Robert W. Hull, *The Origin and Development of Waterway Policy of the United States* (Washington: National Waterways Conference, Inc, 1967)

Hullfish, William R., *The Canaller's Songbook* (York, Pa: The American Canal and Transportation Center, 1984, 1987).
—————————————, "Songs of the 'Horse-Ocean Sailor'," *AC*, Bulletin No. 62 (August 1987), 4-5.

Preliminary report of the Inland Waterways Commission, (60th Congress, 1st Session, Senate Doc. 325, February 26, 1908).

James, Edmund J., "The Canal and the Railway," *American Economic Association Publications,* V (1890), 281-357.

Johnson, Emory R. et al, *History and Domestic Foreign Commerce in the United States,* 2 vols. (Washington: Carnegie Institution, 1915, 1922; reprint, Burt Franklin, n.d.

Johnson, Gilbert R., "United States-Canadian Treaties Affecting Great Lakes Commerca and Navigation," *Inland Seas,* III (1947), 203-207; V (1948), 113-119.

Jones, A., "Cruising the Inland Waterway," *Reader's Digest,* CXXVI (April 1985), 172-179.

Judson, William V., "The Services of graduates as Explorers, Builders of Railways, Canals, Bridges, Light-Houses, Harbors and the Like," in *The Centennial of the United States Military Academy at West Point, New York, 1802-1902,* 2 vols. (Washington: Government Printing Office, 1904), I, 835-873.

Kierans, Thomas W., "The Great Recycling and Northern Development (GRAND) Canal Concept," in Robert Salkeld, Frank P. Davidson and C. Lawrence Meador (eds.), *Macro-Engineering, The Rich Potential* (New York: American Institute of Aeronautics and Astronautics, 1981).
—————————————, "Thinking Big in North America: The Grand Canal Concept," *Futurist,* XIV(6) (1980), 29-32.

Kirby, Richard S., *Inventors and Engineers of Old New Haven* (New Haven, Conn: New Haven Colony Historical Society, 1939), "Early Yale Engineers," 37-54.

Kirkland, Edward C., *Men Cities and Transportation: A Study of New England History, 1820-1900,* 2 vols (Cambridge, Mass: Harvard Univ. Press, 1948), pps. 60-101.

GENERAL WORKS AMERICAN- (cont.)

Larson, John L., "'Bind the Republic Together': The National Union and the Struggle for A System of Internal Improvements," *Journal of American History*, LXXV (1987), 363-387.

Larson, John W., *Essayons: A History of the Detroit District U.S. Army Corps of Engineers* (Detroit, Michigan: U.S. Army Corps of Engineers, Detroit District, 1981).

Latcha, J.A., "Railroads Versus Canals," *North American Review*, CLXVI (1898), 207-225.

Laurent, Jerome K., "Trade, Transport and Technology: the American Great Lakes, 1866-1910," *Journal of Transport History*, 3d series, IV (1983), 1-24.

Lebergott, Stanley, "United States Transport Advance and Externalities," *Journal of Economic History*, XXVI (1966), 437-465.

Lively, Robert A., "The American System, A Review Article," *Business History Review*, XXIX (March 1955), 81-96.

Luxon, Norval N., *Niles' Weekly Register: News Magazine of the Nineteenth Century* (Baton Rouge: Louisiana State University Press, 1947), Ch 10, "Roads, Rivers, Canals and Railroads.

McCausland, Elizabeth, *The Life and Work of Edward Lamson Henry N.A: 1841-1919* (New York State Bulletin No. 339, September 1945), includes canal art.

MacGill Caroline E. *et al*, *History of Transportation in the United States Before 1860* (Washington, D.C: Carnegie Institution, 1917; reprint, Peter Smith, 1948).

McGrane, Reginald C., *Foreign Bondholders and American State Debts* (New York: Macmillan, 1935).

McIlwraith, Thomas F., "Freight Capacity and Utilization of the Erie and Great Lakes Canals before 1850," *Journal of Economic History*, XXXVI (1976), 852-877.

McKelvey, William J., Jr., *Champlain to Chesapeake: A Canal Era Pictoral Cruise* (Exton, Pa: Canal Press, Inc, 1978).
——————————————————, "Researching Canal Boats," *AC* No. 25 (May 1978), 2-3.

McNown, J.S., "Canals in America," *Scientific American*, CCXXXV (July 1976), 116-124.

GENERAL WORKS–AMERICAN (Cont.)

Maass, Arthur, *Muddy Waters: The Army Engineers and the National Rivers* (Cambridge, Mass: Harvard Univ. Press, 1951).

Mayo, Robert S., "Early American Tunnels," *CC*, No. 40 Autumn 1977), 6-8.

Merritt, Raymond H., *Engineering in American Society* (Lexington, Kentucky: Univ. of Kentucky Press, 1969)

Mitchell's Compendium of the Internal Improvements of the United States: Comprising General Notices of All the Most Important Canals and Rail=Roads throughought the Several States and Territories of the Union (Philadelphia, Mitchell & Hinman; 1838, reprint, Glen Echo, Md: American Canal Center, 1972).

Morgan, Arthur B., *Dams and Other Disasters: A Century of the Army Corps of Engineers* (Boston: Porter Sergeant, 1971).

Morison, Elting E., *From Know-How to Nowhere: The Development of American Technology* (New York: Basic Books, 1974); reprint, New York: New American Library, 1977), Chapters I, II and III.

Morris, Charles N., "The Progress of Navigation and Commerce on the Waters of the Mississippi River and the Great Lakes, A.D. 1700 to 1846." *Publications of the Mississippi Historical Society*, VII (1903), 479-523,

Moulton, Harold G., *Waterways versus Railways* (Boston and New York: Houghton Mifflin Co., 1912, 1926).

Musham, H.A., "Ships That Went Down to the Seas," *Inland Seas*, I (October, 1945), 2-13; II (1946), 17-27;

Nelson, E.C., "Presidential Influence on the Policy of Internal Improvements," *Iowa Journal of History and Politics*, IV (1906), 3-69.

Nettles, Curtis, "The Mississippi Valley and the Constitution, 1815-1829," *Mississippi Valley Historical Review*, XI (1924), 332-357.

Newlands, Francis G., "The Use and Development of American Waterways," *Annals of the American Academy of Political and Social Science*, XXXI (1908), 48-66.

GENERAL WORKS- AMERICAN (Cont.)

Niemi, Albert W., Jr., "A Closer Look at Canals and Western Manufacturing in the Canal Age: A Reply," *Explorations in Economic History*, IX (1972), 423-424.

----------------------, "A Further Look at Interregional Canals and Economic Specialization, 1820-1840," *Explorations in Economic History*, VII (1970), 499-520.

----------------------, "Interregional Canals and Manufacturing Development in the West of U.S.A. before 1840," *Revue Internationale d'Historie de la Banque* [Italy]. IX (1974), 192-212.

One Hundred Years' Progress of the United States. . .By Eminent Literary Man (Hartford, Conn: L. Stebbins, 1870, reprinted by Arno Press, 1972), "Coasters-Steamboats-Canals," 178-191.

Parkman, Aubrey, *Army Engineers in New England: The Military and Civil Work of the Corps of Engineers in New England 1775-1975* (Waltham, Mass: U.S. Army Corps of Engineers, New England Division, 1978).

Phillips, Ulrich B., *A History of Transportation in the Eastern Cotton Belt to 1860* (New York: Columbia Univ. Press, 1908, reprint, Octagon, 1968).

Poor, H. V., *History of the Railroads and Canals of the United States*, vol I (New York: J. H. Schultz & Co., 1860); reprint Augustus M. Kelly, 1960).

------------, *Sketch of the Rise and Progress of Internal Improvements and of the Internal Commerce of the United States* (New York: 1881), reprinted from H.V. and H.W. Poor, *Manual of the Railroads of the United States for 1851. . .*(New York: H.V. and H. W. Poor, 1881.)

Powell, Fred W., "First Canals on American Continent," *Journal of American History*, IV (1910), 407-416 [early New England Canals].

Purdy, T.C., Special Agent, "Report on the Canals of the United States," in U.S. Congress, House Misc. Doc 42, vol 13, pt. 4, Department of Interior, Census Office, *Report on the Agencies of Transportation in the United States* (Washington: Govt. Printing Office, 1883), 725-764.

Rae, John Bell, "Federal Land Grants in Aid of Canals," *Journal of Economic History*, IV (1944), 167-177.

Ransdell, Joseph E., "Legislative Program Congres Should Adopt for Improvement of American Waterways," *American Academy of Political and Social Science*, XXXI (1908), 36-47.

GENERAL WORKS—AMERICAN (CONT.)

Ranson, Roger, "Canals and Development: A Discussion of Issues," *American Economic Review*, LIV (1965), 365–376.

——————, "A Closer Look at Canals and Western Manufacturing," *Explorations in Economic History*, VIII (1971), 501–508; "A Reply," see Niemi; Ransom, "A Rebuttal," *Explorations in Economic History*, IX (1972), 425–426.

——————, "Interregional Canals and Economic Specialization in the Antebellum United States," *Explorations in Entrepreneurial History*, 2d series, V (Fall 1967), 12–35.

——————, "Public Canal Investment and the Opening of the Old Northwest," in David C. Klingaman and Richard K. Vedder, (eds.), *Essays in Nineteenth Century Economic History: The Old Northwest* (Athens: Ohio, Ohio Univ. Press, 1975), 246–268.

——————, "Social Returns from Public Transport Investment: A Case Study of the Ohio Canal," *Journal of Political Economy*, LXXV (1970), 1041–1060.

Rapp, Marvin A., "The Niagara Seaway– All American Canal," *Inland Seas*, XXI (1965), 49–58.

"Report of the Franklin Institute on the Inclined Plane of Professor James Renwick," *Journal of the Franklin Institute*, II (1826), 257–263, 321–324; Letter from Renwick in Response, II, 329–332.

"Art. I.–––1. Report of the Examination which has been made by the Board of Engineers, with a view to Internal Improvement, &c. February 14, 1825. . .
    2. Information required by a Resolution of the House of Representatives. . in Relation to Expenditures incident or relating to Internal Improvements, for the Years 1824 and 1825.
    3. Report of the Board of Internal Improvement upon the Subject of a National Road. . .," *North American Review*, n.s., XXIX (1827), 1–13.

"Report on E & T Fairbanks & Co's Weigh Lock for Canals, (with a Plate)," *Journal of the Franklin Institute*, LI (1851), 430–432.

"Art I-1. Report of the Examination which has been made by the Board of Engineers with a view to Internal Improvement, &c. February 14, 1825. Printed, by Order of the Senate.
2. Information required by a Resolution of the House of Representatives of the 13th ult. in Relation to Expenditures incident or elating to Internal Improvements, for the Years 1824 and 1825. Read and laid on the Table, April 3, 1826. . . *North American Review*, new series No. XXIX (January 1827), 1–23.

GENERAL WORKS— AMERICAN (cont.)

Ringwalt, John L., *Development of Transportation Systems in the United States* (Philadelphia: Railway World Office, 1888, Johnson Reprint, 1966).

Roebling, John A., "American Manufacture of Wire Rope, for inclined planes, standing rigging, mines, tillers, &c," *Journal of the Franklin Institute*, XXXVII (1844), 57-60.

Roosevelt, Theodore, "Our National Inland Waterways Policy," *Annals of the American Academy of Political and Social Science*, XXXI (1908), 1-11.

Report on Steam Carriages by a Select Committee of the House of Commons of Great Britain with Minutes of Evidence and Appendix, U. S. Govt., 22d Congress, 1st Session [Doc. No. 101]. House of Representatives, 1-147, with documents to 347, (Washington, Duff Green, 1832).

Sayenga, Donald, "Canals, Converters, and Cheap Steel," *CCHT*, 75-103.

Schaefer, Donald and Thomas Weiss, "The Use of Simulation Techniques in Historical Analysis: Railroads Versis Canals," *Journal of Economic History*, XXXI (1971), 854-884.

Scheiber, Harry N., "Government and the Economy: Studies of the 'Commonwealth' Policy in Nineteenth Century America," *Journal of Interdisciplinary History*, III (Summer 1972), 135-151.
——————————, "On the New Economic History—and Its Limitations: A Review Essay," *Agricultural History*, XLI (1967), 383-395.
——————————, "The Rate-Making Power of the State in the Canal Era," *Political Science Quarterly*, LXXVII (1962), 397-413.

Scheiber, Harry N. and Stephen Salisbury, "Reflections on George Rogers Taylor's *The Transportation Revolution, 1815-1860*: A Twenty-five Year Retrospect," *Business History Review*, LI (1977), 79-89.

Schmidt, Louis B., "Internal Commerce and the Development of the National Economy Before 1860," *Journal of Political Economy*, XLVII (1939), 798-822.

Schodek, Daniel L, *Landmarks in American Civil Engineering* (Cambridge, Mass. and London: MIT Press, 1987), Chapter I, "Canals," 1-27.

GENERAL WORKS, AMERICAN   (Cont.)

Schwantes, Carlos A., "Promoting America's Canals: Popularizing The Hopes and Fears Of The New American Nation," *Journal of American Culture*, I (Winter 1978), 700-712.

Shallat, Todd, "Building Waterways, 1802-1861," *Technology and Culture*, XXXI (Jan 1990), 18-50.

Shank, William H. with Hahn, Mayo and Hobbs, *Towpath to Tugboats* (York, Pa: American Canal and Transportations Center, 1982, 2 ed. 1985).
------------------ (ed.), *The Best from American Canals: Number 1* [1972-1979] (York, Pa: American Canal and Transportation Center, 1980)
------------------ (ed.), *The Best from American Canals: Number 2* [1980-1983] (York, Pa: American canal and Transportation Center, 1984).
------------------ (ed.), *The Best from American Canals: Number 3* [1983-1986] (York, Pa: American Canal and Transportation Center, 1986).
------------------ (ed.), *The Best from American Canals: Number 4* [1986-1989] (York, Pa: American Canal and Transportation Center 1989)

Shaw, Ronald E., *Canals for a Nation: A History of the Canal Era in the United States, 1790-1860* (Lexington: Univ. of Kentucky Press, 1990).

Shelton, William A., "The Lakes-to-the-Gulf Deep Waterway," *Journal of Political Economy*, XX (1912), 541-573, 653-675, 765-806.
-------------------, *The Lakes-to-the-Gulf Deep Waterway*, reprinted with additions from *Journal of Political Economy*, (Chicago, Univ. of Chicago Press, 1912).

Snyder Frank E. and Brian H. Guss, *The District: A History of the Philadelphia District U. S. Army Corps of Engineers, 1866-1971* (Philadelphia, U. S. Army Engineer District, January 1974).

"Specifications of certain Improvements in the Balance Lock, or Inclined Plane intended as a Substitute for Locks on Canals, especially on such canals as have a great lift together with a scarcity of water. Invented by MINUS WARD, Civil Engineer," *Journal of the Franklin Institute*, III (1827), 91-94; "A Further Description of Wards Balance Lock", III, 95-97.

GENERAL WORKS, AMERICAN (CONT.)

Springer, Ethel M., Canal Boat Children," *Monthly Labor Review* XVI, No. 2, February 1923, 227-247, Washington, D.C: U. S. Department of Labor, 1923, reprinted as *Canal Boat Children on the Chesapeake and Ohio, Pennsylvania and New York Canals* with additions by Thomas Hahn (Shephardstown, W. Va: Center for American Canal and Transportation History, 1977)

Stapleton, Darwin H., "The Origin of American Railroad Technology, 1825-1840," *Railroad History* CXXXIX (Autumn 1978), 65-77.
————————————, *The Transfer of Early Industrial Technologies to America* (Philadelphia: American Philosophical Society, 1987).

Stevenson, David, *Sketch of the Civil Engineering in America*, 2d ed., (London: John Weale. 1838, 1859).

Stover, John F., "Canals and Turnpikes: America's Early-Nineteenth-Century Transportation Network," in Joseph R. Frese and Jacob Judd (eds.), *An Emerging Independent American Economy* (Tarrytown, N.Y: Sleepy Hollow Press, 1980), 60-98.

Swanson, Leslie C., *Canals of Mid-America* (Moline, Ill: Swanson Publishing Co., 1969?).

Taylor, George Rogers, *The Transportation Revolution 1815-1860* (New York: Harper & Row, 1951).

Tobin, Catherine, "Irish Labor on American Canals," *CCHT*, IX (1990), 163-186.

Transportation and the Early Nation, Papers Presented at an Indiana American Revolution Bicentennial Symposium, Allen County-Fort Wayne Historical Society Museum, Fort Wayne, Indiana, April 14-26, 1981 (Indianapolis: Indiania Historical Society, 1982. contains the following articles:
    Scheiber, Harry N., "The Transportation Revolution and American Law: Constitutionalism and Public Policy," 1-29;
    Clanin, Doughlas E., "Internal Improvements in National Politics, 1816-1830," 30-60;
    Zimmer, Donald T., "The Ohio River: Pathway to Settlement," 61-88;
    Shaw, Ronald E., "The Canal Era in the Old Northwest," 89-112;
    Gray, Ralph D., The Canal Era in Indiana, 113-134;
    Stover, John F., Iron Roads in the Old Norethwest: Railroads and the Growing Nation, 135-156.

Trout, William E., *The American Canal Guide*, Part 1, British Columbia, Washington, Oregon and California (1974), Part 2, N. & S.Carolina, Georgia and Florida (1975); Part 3, Lower Mississippi, Louisiana, Texas, Arkansas, Tennessee and Alabama (1974) and Part 4, West Virginia, Kentucky and the Ohio River (1988), American Canal Society.

GENERAL WORKS–AMERICAN (cont.)

Watson, Elkanah, *Men and Times of the Revolution*, 2nd ed., Winslow C. Watson, ed.,(New York: Dana and Company, 1857, reprint ed. by Crown Point Point Press, 1968), includes Benjamin Franklin on canals, and Watson on canals in France, Belgium, Netherlands and England, 1779-1784.

Waugh, Robert C., Jr., "Canal Development in Early America," *AC*, part I, No 30 (August 1979), 3-4; part 2, No. 31 (November 1979),4-5,6.

Weaver, S., "Along the Towpath: Packet Rides, Museums, and Restored Villages Chronicle Life on America's Historical Canals," *Americana*, XIV, Nov/Dec 1986), 6+.

Wellons, Charles McCartney, "Construction and Operation of a Modern River Lock," *National Waterways*, VI (January 1929), 25-28, 57-59.

Willoughby, William R., "Early American Interest in Waterway Connection Between the East and the West," *Indiana Magazine of History*, LII (1956), 319-342.

Woolfolk, George R., "Rival Urban Communication Schemes for the Possession of the Northwest Trade, 1783-1800," *Mid-America*, XXXVIII (1956), 214-232.

JUVENILE

Boyer, Edward, *River and Canal* (New York: Holiday House, 1986)

Franchere, Ruth, *Westward by Canal* (New York: Macmillan, 1972).

Richard, G., *Canals* (New York: Bookwright, 1988).

GENERAL WORKS—AMERICAN (CONT.

Unpublished

 Abbott, Frederick K., "The Role of the Civil Engineer in Internal Improvements: The Contributions of the Two Laommi Baldwins Father and Son, 1776-1838," PhD, Columbia, 1952.

 Formwalt, Lee W., "Benjamin Henry Latrobe and the Development of Internal Improvements in the New Republic, 1796-1820," PhD, Catholic University, 1977.

 Harrison, Joseph A., "The Internal Improvement Issue in the Politics of the Union, 1783-1825," PhD, Virginia, 1954.

 Isard, Walter, "The Economic Dynamics of Transport Technology," PhD. Harvard, 1943.

 Sandove, Abraham H., "Transport Improvement and the Allegheny Barrier: a Cost Study in Economic Innovation," PhD, Harvard, 1928.

 Segal, Harvey H., "Canal Cycles, 1834-1861: Public Construction Experience in New York, Pennsylvania, and Ohio," PhD, Columbia, 1956.

 Shallot, Todd A., "Structures in the Stream: A History of Water, Science and the Civil Activities of the U.S. Army Corps of Engineers," PhD, Carnegie-Mellon, 1985.

BIOGRAPHIES

Bobbe, Dorothie, *DeWitt Clinton* (New York, Minton, Balch & Co., 1933).

Boucher, Cyril T.G., *John Rennie, 1761-1821: The Life and Work of a Great Engineer* (1963, reprint, New York, Augustus M. Kelley, 1967).

Ferdinand de Lesseps
   Beatty, Charles, *DeLesseps of Suez* (New York, Harper, 1956).
   Burlingame, Roger, "The 'Great' Frenchman," *Scribner's Magazine* XCIV (1933), 208-213.
   Schoenfield, Hugh J., *Ferdinand deLesseps* (London, Herbert Joseph Ltd., 1937)
   Smith, G. Barrett, *The Life and Enterprises of Ferdinand de Lesseps*, (London, W. H. Allen & Co. Ltd., 1895).

Dorwart, Harold L., "Biographical Notes on Jonathan Knight" (1787-1858), *PMHB*, LXXV (1951), 76-89.

Oliver Evans:
   Bathe, Grenville and Dorothy, *Oliver Evans: A Chronicle of Early American Engineering*, (Philadelphia, Historical Society of Pennsylvania, 1935, ARNO reprint, 1972).
   Ferguson, Eugene, *Oliver Evans: Inventive Genius of the American Industrial Revolution* (Greenville, Del, Hagley Museum, 1980).

Fitzsimons, Neil (ed.), *Engineer As Historian* (Kensington, Maryland, P.J. Lorimore, 1985).

Robert Fulton:
   Colden, Cadwallader David, *The Life of Robert Fulton* (New York, , 1817).
   Dickinson, H.W., *Robert Fulton, Engineer and Artist: His Life and Works* (London, John Lane, 1913).
   Knox, Thomas W., *The Life of Robert Fulton* (N.Y., G. P. Putnam's Sons, 1886).
   Phillip, Cynthis Owen, *Robert Fulton, A Biography* (N.Y., Franklin Watts, 1985).
   Sutcliffe, Alice Crary, *Robert Fulton and the Clermont* (N.Y., The Century Company, 1909).

BIOGRAPHY (cont.)

 Gilchrist, Agnes Addison, *William Strickland, Architect and Engineer* (Philadelphia, Univ. of Pennsylvania Press, 1950).

George W. Goethels
 Bishop, Joseph Bucklin, "Personality of Colonel Goethals," *Scribner's Magazine* LVII (1915).
 Bishop, Jospeh Bucklin and Farnham Bishop, *Goethals, Genius of the Panama Canal: A Biography* (New York and London, Harper & Brothers, 1930).

 Harrison, Joseph H, Jr., Simon Bernard, the American System and the Ghost of the French Alliance," in John B. Boles (ed.), *America The Middle Period, Essays in Honor of Bernard Mayo* (Charlottsville, University Press of Virginia, 1973), 145-167.

 Hobbs, T. Gibson, Jr., "Edward Hall Gill, Civil Engineer," *CC*, No. 49 (Winter 1980), 15.

Benjamin Henry Latrobe
 Benson, Barbara, (ed.), *Benjamin Henry Latrobe and Moncure Robinson: The Engineer as Agent of Technological Transfer* (Greenville, Del., Eleutherian Mills Historical Library, 1975).
 Carter, Edward C., II et al (editors), *The Journals of Benjamin Henry Latrobe, 1794-1820*, 3 vols (New Haven, Published for the Maryland Historical Society by Yale Univ. Press, 1977, 1980.
 Hamlin, Talbot, *Benjamin Henry Latrobe* (New York, Oxford Univ. Press)
 Stapleton, Darwin H. (ed.), *The Engineering Drawings of Benjamin Henry Latrobe (Latrobe Papers,* Series II, Vol. I), (New Haven, Yale University Press for Maryland Historical Society, 1980).
 VanHorne, John C. et al (editors), *The Correspondence and Miscellaneous Papers of Benjamin Henry Latrobe*, 3 vols to date, (New Haven and London, published for the Maryland Historical Society by the Yale Univ. Press, 1984-1988.

 Lewis, Gene D., *Charles Ellet, Jr.: The Engineer as Individualist, 1810-1862* (Urbana, Univ. of Illinois Press, 1968).

Frederick List
 Altner, Hanns G., "Frederick List (1789-1846), The City of Reading, His Stepping Stone to Fame and Greatness, An Apotheosis," *Historical Review of Berks County*, I, No 1 (Oct. 1935), 6-11.
 Bell, John F., "Frederick List, Champion of Industrial Capitalism," *PMHB*, LXVI (1942), 56-83.
 Henderson, W.O., *Friedrich List: Economist and Visionary, 1789-1846* (London, Frank Cass, 1983).
 Patton, Spiro G., "Frederick List's Contribution to the Anthracite/Railroad Connection in the United States," *CCHT*, IX (1990), 3-19.

BIOGRAPHY (cont.)

   David Rittenhouse
      Fox, Edward, *David Rittenhouse* (Philadelphia, Univ. of
         Pennsylvania, 1946)
      Hindle, Brooke, *David Rittenhouse* (Princeton, Princeton Univ.
         Press, 1964).

   Moncure Robinson
      Osborne, Richard Boyse, *Professional Biography of Moncure Robinson,
         Civil Engineer* (Philadelphia, J.B. Lippincott Co., 1889).
      see Benson under Latrobe listing.

      Rowe, Kenneth W., *Matthew Carey: a Study in American Economic
         Development*, Johns Hopkins Studies in History and Political
         Science, LI (4), (Baltimore, 1933).

      Stuart, Charles B., *Lives and Works of Civil and Military Engineers
         of America* (New York, D. Van Nostraand, 1871).

      "Obituary, John C. Trautwine," *Journal of the Franklin Institute*,
         CXVI (1883), 390-396.

      Turnbull, Archibald Douglas, *John Stevens: An American Record*
         (N.Y.,The Century Company, 1928).

      Vose, George L., *A Sketch of the Life and Works of Loammi Baldwin .
         . . .* (Boston, Press of G.H. Ellis, 1885).

   George Washington Whistler
      Parry, Albert, *Whistler's Father* (Indianapolis and New York, Bobbs
         Merrill Co, 1939).
      Vose, George L., *The Life and Works of George Washington Whistler*
         (Boston, Lee and Shepherd, New York, O.T. Dillingham, 1887).
      ---------------, "George W. Whistler, C.E.," *Journal of the
         Association of Engineering Societies*, VI (1886-1887), 37-52.

      Ward, James A., *J. Edgar Thomson: Master of the Pennsylvania*
         (Westport, Conn, Greenwood Press, 1980).

BIOGRAPHIES (Cont.)
    Josiah White
        Richardson, Richard, *Memoir of Josiah White*, Philadelphia: J.B. Lippencott & Co., 1873).
        Roberts, Solomon W., *Memoir of Josiah White: Written as a Chapter in the History of the Lehigh Valley* (Easton, Pa: Boxler & Corwin, 1860).
        Josiah White's History given by himself, (Philadelphia: Press of G.H. Buchanan Co., 1909), reprinted by Carbon County Board of Commissioners, June 29, 1979

White, William Pierrepont, "Canvass White's Services," *Buffalo Historical Society Publications,* XIII (1909), 353-366.

Wood, Richard G., *Stephen Harriman Long 1784-1864* (Glendale, California: Arthur H. Clark Co., 1966).

PENNSYLVANIA

Bibliographical Aids

Cummings, Hubertis M. (compiler), *Pennsylvania Board of Canal Commissioners' Records with Allied Records of Canal Companies Chartered by the Commonwealth-- Descriptive Index* (Harrisburg: Bureau of Land Records, Department of Internal Affairs, 1959)

for additional description see
Cummings, Hubertis M., "James D. Harris, Canal Engineer: Notes on His Papers and Related Canal Papers," *PH*, XVIII (1951), 31-45.
Heydinger, Earl, Review in *PH*, XXVII (1960), 326-8

Hasse, Adelaide R. *Index of Economic Material in Documents of the States: Pennsylvania, 1790-1904*, 3 vols, (Washington: Carnegie Insitution, 1919, reprint, Kraus Reprint, 1965) "Canals and Slack-Water Navigation," I, 379-485; "Commerce; Canal-borne," I, 595-597.

Simonetti, Martha L (comp.), *Inventory of Canal Commissioners' Maps in the Pennsylvania State Archives*: RG 17 Records of the Land Office: D. Board of Canal Commissioners Map Books 1810-1881, (Harrisburg, Pa: Pennsylvania Historical and Museum Commission Bureau of Archives and History, 1968).

Simonetti, Martha L. (compiler) and Donald H. Kent and Harry E. Whipkey (eds.), *Descriptive List of the Map Collection in the Pennsylvania State Archives* (Harrisburg, Penna: Historical and Museum Ccmmission, 1976)

Other works

Albig, W. Espy, "Early Development of Transportation on the Monongahela River," *WPHM*, II (1919), 115-124.

Allegheny Portage Railroad: Its Place in the Main Line of Public Works of Pennsylvania, Forerunner of the Pennsylvaqnia Railroad (Pennsylvania Railroad Information, II No.1, February 1930).

[Allegheny Portage Railroad], "'Old Portage' Marked Important Step in State's Transportation," *PDIA*, VI, #6 (1938), 13-28.
――――――――――――――――――――――, "Tests on Old Portage Led to Use of Locomotives Instead of Horses," *PDIA*, VII, # 1 (1939, 16-31.
――――――――――――――――――――――, "Building of New Line Paved Way for Passing of Old Portage Road," *PDIA*, VII, # 2 (1939), 21-29.

PENNSYLVANIA (Cont)

"The Allegheny River," from the U.S. Army Corps of Engineers Navigation Chart Book, 1976, *CC*, No. 57 (Winter 1982), 3, 15.

Allen, Charles, *Second Geological Survey of Pennsylvania, 1875-7* (Harrisburg: 1878).

Anderson, George W., "Philadelphia to Lewisburg in 1846," from *Christian Chronicle*, Oct. 28, 1846, *CC*, No. 17 (Summer, 1971), 6.

Anderson, John A., "Navigation on the Delaware and Lehigh Rivers," *Bucks County Historical Society Papers*, IV (1917), 282-312.

"Ann Royal Revisited," *CC* No.48 (Autumn 1979), 14.

Archer, Robert F., *A History of the Lehigh Valley Railroad: "The Route of the Black Diamond"* (Berkeley, California: Howell-North Books, 1977). [Chapters I and II].

Armor, William C., *Lives of the Governors of Pennsylvania* (Norwich, Conn: T. H. Davis, 1874).

Armroyd, George, *A Connected View of the Whole Internal Navigation of the United States*, (Philadelphia: Carey & Lea, 1824; revised, author, 1830)

Aungst, Dean M., "The Schuylkill and Susquehanna Canal," *CC*, No. 4 (Spring 1968), 3-4.
————————, "Two Canals of Lebanon County," *The Lebanon County Historical Society*, XIV, No. 1, 1966.
————————, *The Union Canal and the Lehmans* (Lebanon, Pa: Lebanon County Historical Society, 1985).

Backofen, Catherine, "Congressman Harmar Denny," *WPHM*, XXIII (1940), 65-78.

Bailey, Rev. Richard S., "Night Boat from Philadelphia," *AC*, No. 46 (August 1983), 8-9.

Baldwin, Leland D. (compiler), "Charles Dickens in Western Pennsylvania," *WPHM*, XIX (1936), 27-46.

Barber, David G., *A Towpath Guide to the Lehigh Canal, Lower Division* (Delaware Valley Chapter, Appalachian Mountain Club, 1981).

PENNSYLVANIA (Cont.)

Barker, Charles R., "Philadelphia, 1836-1839, Transportation and Development," *Philadelphia City History Society Publications*, II (1933), 337-370.

Barnes, Horace B., "Organization and Early History of the Conestoga Navigation Company," *Papers of the Lancaster County Historical Society*, XXXIX (1935), 49-61.

Bartholomew, Ann and Lance Metz, *The Delaware and Lehigh Canals* Easton, Penna: Center for Canal History and Technology, 1988).

Baumgardner, Mahlon J., and Floyd C. Hoenstine, *The Old Allegheny Portage Railroad* (Ebensburg: Blair County Historical Society, 1952).

Baumgardner, Mahlon J., "The Summit Hotel Formerly Known as The Summit Mansion House," *PDIA*, XX, # 12 (1952), 3-7, 20-1.

Bell, John F., "Robert Fulton and the Pennsylvania Canals," *PH*, IX (1942), 191-196.

Benner, Dorothea O., "The D and H Canal, Pennsylvania and New York," *New York Folklore Quarterly*, VI (Winter, 1950), 260-267; Chapter VI from 1950 Cornell University thesis entitled "Patch by Patch, Lore of Wayne County, Pennsylvania."

Best, Gerald M. "The Gravity Railroad of the Delaware and Hudson Canal Company," *Railway and Locomotive Historical Society*, Bulletin # 82, (April 1951), 7-24.

Binder, Frederick M., *Coal Age Empire: Pennsylvania Coal and its Utilization to 1860*, (Harrisburg: Pennsylvania Historical and Museum Commission, 1974).

Bishop, Alvard L. "Corrupt Practices Connected with the Building and Operation of the State Works of Pennsylvania," *Yale Review*, XV (1907), 391-411.
---------------- "The State Works of Pennsylvania," *Transactions of the Connecticut Academy of Arts and Sciences*, XIII (Nov. 1907), 149-297.

Bothwell, Margaret, "Incline Planes and People: Some Past and Present Ones," *WPHM*, XLVI (1963), 311-346.

Bowen, Eli, *The Pictorial Sketch Book of Pennsylvania* [1st edition Philadelphia: Hazard, 1852], [2d. ed., Phila: W.Bromwell, 1853]. [eighth edition, "revised and greatly enlarged," Wm. White Smith, 1854].

PENNSYLVANIA (Cont.)

Bowman, John B., "Schuylkill Canal Memoirs," *CC*, No 24 (Spring 1973), 3-4; No. 25 (Summer 1973), 3-4; No. 26 (Fall 1973), 3-4; No. 27 (Winter 1974), 3-4; No. 28 (Spring 1974), 3-4; No. 29 (Winter 1974), 3-4; No. 30 (Fall 1974), 3-4; No. 31 (Winter 1975), 3-4; No. 35 (Spring 1976), 11.

—————————, "The Schuylkill Rangers," *Pennsylvania Folklife*, IX, (No. 1, Winter, 1957-58), 18-23.

—————————, "The Schuylkill Rangers," *Schuylkill County Historical Society Publications*, VI (1950), 41-50.

"The Diary of Samuel Breck, 1827-1833," *PMHB*, CIII (1979), 222-251, 357-382, 497-527.

Breck, Samuel, *Sketch of the Internal Improvements Already Made by Pennsylvania . . .* (Philadelphia: J. Maxwell, 1818, 2d. revised and enlarged, M. Thomas, 1818).

Brenner, Fred J., *Hiking With History Along the Erie Extension Canal*, (Mercer, Pa: Mercer County Promotion Agency, 1967).

Brewer, George S., "Transportation in Erie County," *The Journal of Erie Studies*, VI, No.1 (1977), 44-59.

Brown, Revelle Wilson, "Pennsylvania's Greatest Builder of Railroads [Moncure Robinson]," *PDIA*, XVIII, # 1 (1947), 3-4, 20-27.

Buckingham, J.S., *The Eastern and Western States of America*, 3 volumes, (London: Fisher, Son, & Co., 1842).

Bugbee, Leroy, "The North Branch Canal," *Proceedings and Collections of the Wyoming Historical and Geological Society*, XXIV (1984).

"Building of New Line Paved Way for Passing of Old Portage Road," *PDIA*, VII # 2 (1939), 2, 21-29.

Burgess, George H. and Miles C. Kennedy, *Centennial History of the Pennsylvania Railroad*, (Philadelphia: Pennsylvania Railroad, 1949).

"Canals and Churches Financed by Lotteries in Pennsylvania," *PDIA*, XVII No. 12 (1949), 27-29.

"Canal Boat Seizures- Continued," *CC*, No. 72 (Fall 1985), 3-4, 12.

"Canal Building Proved Helpful in the Development of Industry," *PDIA*, VII, # 5 (1939), 10-23.

"Canal Closing and Lawsuits," *CC*, No 74 (Spring 1986), 11-16.

PENNSYLVANIA (Cont)

"Canal Crisis--Federal Seizure of Boats," *CC*, No. 70 (Spring 1985), 13-16.

"Canalization as Early Means of Transportation Conceived by Penn," *PDIA*, III, # 5, 2-8.

Carey, Matthew, *Brief View of the System of Internal Improvement of the State of Pennsylvania* original edition (Lydia R. Baily, June 13th 1831), 2d edition (Philadelphia: Lydia R. Baily, July 9, 1831).

Carlson, W. Bernard, "The Pennsylvania Society for the Promotion of Internal Improvements: A Case Study in the Political Uses of Technological Knowledge, 1824-1825," *CCHT*, VII 1988), 175-206.

Carlson, Robert E., "The Pennsylvania Improvement Society and its Promotion of Canals and Railroads, 1824-1826," *PH*, XXXI (1964), 295-310.

Case, Lynn M., "Philadelphia and Baltimore in 1837, Notes written by an Italian Count during his journey across America," *PMHB*, LVII (1933), 181-186.

Cassel, Clara M., "Canal Route Rivalries Mark Transportation Pioneering," *PDIA*, VII, # 6 (1939), 4-13.
----------------, "West Branch Canal Route Clash Results in Community Rivalry," *PDIA*, VII, # 7 (1939, 5-9.
----------------, "Canal Construction Nearly Bankrupted Commonwealth," *PDIA*. VII, # 8 (1939), 10-14.
----------------, "Canal Routes Subject of Bitter Battles," *PDIA*, VII, # 9 (1939), 14-17.
----------------, "Lewisburg Canal Cross-Cut Opened 1834," *PDIA*, VII, # 10 (1939), 17-19.
----------------, "Dream of Navigable Route to West is Shattered," *PDIA*, VII, # 11 (1939), 8-11.
----------------, "Canal Construction Problems Revealed by Contracts," *PDIA*, VII # 12 (1939), 19.
----------------, "Canal Construction Contracts Show Comparative Costs," *PDIA*, VII, # 1 (1939), 19-22.
----------------, "Canal Construction Followed By Many Damage Claims," *PDIA*, VIII, # 2 (1940), 8-10.
----------------, "Canal Construction Marked by Lively Happenings," *PDIA*, VIII # 4 (1940), 11-14.
----------------, "Friction and Struggle Mark Canal Construction," *PDIA*, VIII # 5 (1940), 26-28.
----------------, "Muncy Canal Opening Marred by Floods and Accidents," *PDIA*, VIII, # 6 (1940), 16-18.

PENNSYLVANIA (Cont.)

Cassel, Clara M., "Waterman and Washouts Costly to West Branch Canal," DIA, VIII, # 7 (1940), 24-26,
—————————, "Wharves, Basins and Bridges Canal Problems," PDIA, VIII, # 8 (1940), 25-28.
—————————, "West Branch Canal Officials Hear Many Complaints," PDIA, VIII, # 9 (1940), 23-26.
—————————, "West Branch Canal Claims For Damages Heavy," PDIA, VIII, # 10 (1940), 28-29.
—————————,"West Branch Canal Officials Hear Many Complaints," PDIA, VIII # 11 (1940).

Charles, Edwin, "Canal Lore: Conditions Leading to the Building of Canals in Central Pennsylvania," *Pennsylvania German*, XII (1911), 385-394 (also *Snyder County Historical Society Bulletin*, I 1922), 231-239.
—————————, "Canals and Canal Lore," *Proceedings of the Northumberland Society*, II, (1930), 3-132,

Clark, Sara Maynard, "Houseboating on the Canal," *Bucks County Traveler*, IX (Aug 1958), 26-27, 47.

"Conestoga Navigation Company," HR, XI (1833), 254-256.

Coleman, Ernest H,, "Bald Eagle and Spring Creek Navigation," CC, No. 19 (Winter 1972), 5-6.
—————————,."Philip J. Hoffmann--A Second Look," CC, No. 29 (Summer 1974), 1-2.
—————————, "The Six Clark's Ferry Bridges," AC, No. 44 (February 1983), 4,5.
—————————, "Western Division Canal Boomed Salt Sales," CC, No. 17 (Summer 1971), 3-4.

Corkan, Lloyd A. M., "The Beaver and Lake Erie Canal," WPHS. XVII (1934), 175-188.

Coughtry, W. J., "Construction of the Delaware and Hudson Canal," *The Delaware and Hudson Railroad Bulletin*, X (July 15, 1930), 219-20; (Aug 1, 1930), 229-30, 238; (August 15, 1930) 245-46, 252; (September 1, 1930), 267-68.

Coxe, Tench, *A View of the United States of America in a Series of Papers written at Various Times, in the Years between 1787 and 1794*, [1794], (reprint, Augustus M. Kelley, 1965).

Crippen, Lee F., *Simon Cameron, Ante-Bellum Years*, (Oxford, Ohio: The Mississippi Valley Press, 1942).

PENNSYLVANIA (Cont.)

Cummings, Hubertis M., "Bureau of Land Records' Files Disclose Financial Troubles of Erie Extension Canal Builders," *PDIA*, XXVI, # 4 (1958), 23-29.

----------------------, "Collapse of High Aqueduct and Competition by Railroad Ended Erie Extension Canal's Career," *PDIA*, XXVI, # 6 (1958), 17-22, 27.

----------------------, "History Comes to Life In Revival of Plan for Pittsburgh--Erie Canal," *PDIA*, XXVI # 2 (1958), 1-6, 28-9.

----------------------, "James D. Harris, Canal Engineer: Notes on his Papers and Related Canal Papers" *PH*, XVIII (1951), 31-45.

----------------------, "James D. Harris, Principal Engineer and James S. Stevenson, Canal Commissioner," *PH*, XVIII (1951), 293-306.

----------------------., "James D. Harris and William B. Foster, Jr., Canal Engineers," *PH*, XXIV (1957), 191-206.

----------------------, "John Augustus Roebling, and the Public Works of Pennsylvania," *PDIA*, XXVIII # 5 (1960), 5, 23-4, 32; # 6, 24-27; # 7, 22-25; # 8, 26-29; # 9 (1960), 24-26.

----------------------, "Pennsylvania: Network of Canal Ports," *PH*, XXI (1954), 260-273.

----------------------, "Polish Nobleman's Map of 1824 Halted Visionary Canal Tunnel; Now a Land Records' Treasure, [Charles Treziyulny]," *PDIA*, XXVI, # 9 (1958), 13-15, 28-29.

----------------------, "Some Notes on the State-Owned Philadelphia and Columbia Railroad," *PH*, XVII (1950), 39-49.

Daniel, Warren J., "Pennsylvania Public Improvements Canals and Railroads," *PDIA*. XV, # 7 (1946), 31-32.

Davenport, Frank, "Early Anthracite Events," *CC*, Part 1, No. 65 (Winter 1984), 11-16; Part II, No. 66 (Spring 1984), 14-16.

Davis, Sidney, "The West Branch Canal," *Proceedings of the Northampton County Historical Society*, XXVI (1974), 28-43.

Day, Sherman, *Historical Collections of the State of Pennsylvania*, (Philadelphia: Groton, 1843, reprinted, Port Washington, N.Y: I. J. Friedman, 1969).

Decker, Cora Louise (ed.), "The Log of the Good Ship Molly Polly Chunker," *Bucks County Panaroma*, II, (Sept. 1969), 4-5, 14-15, (Oct. 1969), 4-5, 29-30.

DeGeorge, Theresa Grzankowski, "The Economic Development of the Port of Erie and Related Avenues of Trade: 1825-1845," *The Journal of Erie Studies*, IV, No 1 (1975).

"Delaware Canal: Early History of Waterway From Original Documents," *PDIA*, XXI, # 9 (1953), 3-8, 30; XXI, # 10, 5-20, 32.

PENNSYLVANIA (Cont.)

Delaware and Hudson Company, *A Century of Progress: History of the Delaware and Hudson Canal, 1823-1923* (Albany: J.B. Lyon Co., 1925).

Delaware Canal Master Plan: A Plan and Program to Preserve and Improve The Delaware Canal and Roosevelt State Park [July 1987] (Point Pleasant, Pa: Friends of the Delaware Canal, 1987).

"Diary of Horatio Allen: 1828 (England)," *Bulletin, Railway & Locomotive Historical Society*, LXXXIX (1953), 97-138.

Dickens, Charles, *American Notes for General Circulation*, 2 vols (first edition, London: 1842), many later editions.

"Dickinson [Catherine] Describes the Main Line Trip, 1853," *CC*, No. 70 (Spring 1985), 3-9.

Dickinson, Emily, "By Packet and Rail to Philadelphia [from a letter, June 1856]," *CC*, No. 58 (Spring 1982, 6.

DiMara, John, "Highway Construction Uncovers Canal Remnants," *CC*, No. 74 (Spring 1986), 9-10.

Dinklelacker, Thomas, "The Construction of the Lehigh Canal and the Early Development of the Lehigh Valley Region," *Proceedings of the Wyoming Historical and Geological Society*, XXIV (1984).

Downie, James Vale, "Mule Power Plus Cat Power in Beaver Valley," from *Geneva College Alumnus Magazine, CC,* No. 55 (Summer 1981), 16.

Duane, William J., *Letters Addressed to the People of Pennsylvania respecting The Internal Improvement of the Commonwealth by means of Roads and Canals,* (Philadelphia: Jane Aitken, 1811; reprint, vol XXII, *American Classics in History and the Social Sciences;* reprints, Augustus M. Kelley, 1964; New York: Burt Franklin Press, 1967).
————————, *Observations on the Importance of Improving the Navigation of the River Schuylkill for the Purpose of Connecting it with the Susquehanna* (Philadelphia: n.p., 1818).

Dwyer, Eddie, *A Trip Into Yesteryear and A Tale of Grandpa's Life Aboard a Canal Boat,* (Havre de Grace, Maryland: *Havre de Grace Record*, n.d.), [Susquehanna River and Canals].

PENNSYLVANIA (Cont.)

    Dzombak, William, "Canal Delivers First Ore at Birth of U. S. Steel Industry," *CC*, No. 64 (Autumn 1983), 10, 14.

    —————————, " Canal or Railroad in 1825?" *CC*, No. 78 (Spring 1987), 12, 16.

    —————————, "Canal Voyage Journal–1849," *CC*, No. 78 (Spring 1987), 13–15

    —————————, "The Johnstown Connection," *CC*, No. 67 (Winter 1984), 4–5, 13.

    —————————, "Main Line Canal Hauls Hoosier Canal Funds, 1834," *CC*, No. 74 (Spring 1986), 7–9,16.

    —————————, "Packet Boat Trips Revisited," *CC*, No. 74 (Spring 1986), 1–5.

    —————————, "Steamboats on the Mainline?" *CC*, No. 53 (Winter 1981), 16.

    ————————— (contributor), "Summer Heat––The Erie Canal," [by Tyrone Power] *CC*, No. 79 (Summer 1987), 11–12–16.

    ————————— "Towpath Trail Planned," *CC*, No. 49 (Winter 1980), 5.

    ————————— (contributor), "U.S. Grant Recalls Canal Trip," *CC*, No. 68 (Autumn 1984), 2.

    —————————, "Western Division Sesquicentennial Year!" *CC*, No. 48 (Autumn 1979), 11–13.

"Early Conestoga Navigation," *Papers Read before the Lancaster County Historical Society*, XII (1908), 315–330.

Economic Development Council of Northeastern Pennsylvania, *Transportation in Northeastern Pennsylvania* (Avoca, Penna: Economic Development Council of Northeastern Pennsylvania, 1979), Ch II, "Rivers and Canals."

Ellers, Richard, "Great Lakes Inland Waterway," *AC*, Bulletin No. 53 (May 1985), 8–9, from *Cleveland Plain Dealer*, March 3, 1985 [proposed Canal- Great Lakes, Youngstown, Beaver Falls].

Ellis, Franklin and Samuel Evans, *History of Lancaster County, Pennsylvania, with Biographical Sketches of many of its Pioneers and Prominent Men.*, (Philadelphia: Everts & Peck, 1883).

Ellsworth, fanny, "A Canal That Made History [D&H]," *Travel*, LXXXIII (June 1944), 18–20, 32.

Evans, Henry O., "Notes on Pittsburgh Transportation Prior to 1890," *WPHM*, XXIV (1941), 161–182.

PENNSYLVANIA (Cont.)

Fackenthal, B.F., Jr., "Improving Navigation on the Delaware River with Some Account of Ferries, Bridges, Canals and Floods," *The Bucks County Historical Society Papers*, VI (1932), 103-230.

——————————————, *Improving Navigation on the Delaware River with Some Account of its Ferries, Bridges, Canals and Floods* (Williamsport, Pa: 1927).

——————————————, "Manufacture of Hydraulic Cement in Bucks County," he Bucks County Historical Society Papers, VI (1932), 346-355.

Facts and Arguments in favour of adopting Railways in preference to Canals in the State of Pennsylvania, 4th ed. (Philadelphia: William Fry, Printer, 1825, Arno Reprint in *The Rise Of Urban America*, 1970).

Ferguson, Eugene S. (ed.), *Early Engineering Reminiscences (1815-1840) of George Escol Sellers* (Washington, D.C: Smithsonian Institution, 1965).

Filipelli, Ronald, "Pottsville Boom Town; the Impact of the Schuylkill Navigation Company on the Growth of Pottsville," *Historical Review of Berks County*, XXXV (1970), 126-129.

The First Annual Report of the Acting Committee of the Society for the Promotion of Internal Improvements in the Commonwealth of Pennsylvania, (Philadelphia: 1826).

Fisher, Barbara," Maritime History of Reading, 1833-1905," *PMHB*, LXXXVI (1962), 160-180.

Fisher, Vincent, "A Tale of Two Towns," *CC*, No. 64 (Autumn 1983), 11-14.

"Folklore on the Susquehanna Canals," *CC*, No. 21 (Summer 1972), 3-4 ++.

Folsom, Burton W., Jr., *Urban Capitalists, Entrepreneurs and City Growth in Pennsylavania's Lackawanna and Lehigh Regions, 1800-1920*, (Baltimore: Johns Hopkins Univ. Press, 1981).

Formwalt, Lee W., "Benjamin Henry Latrobe and the Revival of the Gallatin Plan of 1808,", *PH*, XLVIII (1981), 99-128.

Foster, John J., "The Oldest Canal Tunnel in the United States." *PH*, XIX (1952), 352-354.

Foulke, Arthur Toye, "'Ol' Canal Days at Danville," *CC*, No. 78 (Spring 1987), 1, 4-6.

PENNSYLVANIA (Cont.)

Francis, J. G., "The Union Canal," *Lebanon County Historical Society,* II (1939), 225-286.

"French Creek Feeder in the News in 1833," *CC,* No. 43 (Summer 1978), 10-11.

Gaines, Stanley,"Forgotten Highway of Transportation [D & H]," *PDIA,* XVII, # 7 (1949), 3-9; XVII, # 8, 8-11; XVII, # 9, 3-10; XVII, # 10, 9-14; XVII, # 11, 3-6.

Gausler, W.H., "Reminiscences of the Lehigh and Delaware Canal from 1840 to 1856," *The Penn Germania,* I, No. 6 (old series XIII), 452-456; reprinted in American Canals, No. 51 (November 1984), 6-7; No. 52 (February 1985), 4-5.

Gibbons, Edward S., "The Building of the Schuylkill Navigation System, 1815-1828," *PH,* LVII (1990), 13-43.

Giles, Earl B. (contributor), "'Main Line' Sold to P.R.R. June 25, 1857," reproduced from clipping from Ebensburg (Pa.) *Democrat and Sentinel,* July 1, 1857, *CC,* No. 17 (Summer 1971), 5..

Giles, Earl B. and Ralph Michaels (contributors), "'Fire Bug' Destroys Freeport Aqueduct--1848," [from *Apalachian* published in Blairsville, May 17, 1848], *CC,* No. 16 (Spring 1971), 5.

Gillingham, Harold E., "Lotteries in Pennsylvania Prior to 1776," *PH,* V (1938), 77-100.

[Godfrey, Captain Frank H.] XYZ, "West Branch Remembrances," *CC,* No. 24 (Spring 1971), 2.

Gordon, Thomas F., *A Gazetteer of the State of Pennsylvania,* (Philadelphia: T. Belknap, 1833)

Govan, Thomas P., *Nicholas Biddle, Nationalist and Public Banker, 1786-1844,* (Chicago: Univ. of Chicago Press, 1959).

Graeff, Arthur D., "Building Canals in Berks," *Historical Review of Berks County,* XXXIII (1968), 78-81, 130-131, 140-144; XXXIV (1968-1969), 22-3, 28-29.

Alex. Graydon to Jedediah Morse, March 5, 1789, *PMHB,* VI (1882), 114-117.

"Guide to Canal Historical Markers in Pennsylvania," *CC,* No. 52 (Autumn 1980), 7-10.

PENNSYLVANIA (Cont.)

Gurney, Joseph John, *A Journey in North America described in Familiar Letters to Amelia Opie* (Norwich: private printing by Josiah Fletcher, 1841), Letter II, 10-17; partially reprinted as "Across Pennsylvania on a Packet Boat Trip, 1841," *CC* No. 74 (Spring 1986), 5-7, 16.

Gustorf, Fred and Gisela, *The Uncorrupted Heart: Journal and Letters of Frederick Julius Gustorf 1800-1845* (Columbia, Mo: University of Missouri Press, 1969).

Hain, H. H., *History of Perry County, Pennsylvania. . .* (Harrisburg, Pa: Hain-Moore Company, 1922), "Coming of the Canal," 307-420..

Hager, Charles V., *Early History of the Falls of the Delaware, Manayunk, Schuylkill and Lehigh Navigation, Fairmount Waterworks. . .* (Philadelphia: Caxton, Remsen and Haffelfinger, 1869).

Hammond, J.W., *A Tabular View of the Financial Affairs of Pennsylvania From the Commencement of Her Public Works to the Present Time. .* (Philadelphia: Edward Biddle, 1844).

Hare, Jay V., *History of the Reading: The Collected Articles of Jay V. Hare* (Philadelphia: John Henry Strock, 1966.

Hartman, E. J., "Josiah White and the Lehigh Canal," *PH*, VII (1940), 225-235

Hartman, Jesse L., "John Dougherty and the Rise of the Section Boat System,", *PMHB*, CXIX (1945), 294-314.
---------------, "Pennsylvania's Grand Plan of Post-Revolutionary Internal Improvement," *PMHB*, LXV (1941), 437-57.
---------------, "The Portage Railroad National Historic Site and the Johnstown Flood (National) Memorial," *PH*, XXXI (1964), 139-156.

Hartz, Louis, *Economic Policy and Democratic Thought: Pennsylvania, 1776-1860*, (Cambridge: Harvard University Press, 1948).
------------, "Laissez Faire Thought in Pennsylvania, 1776-1860," *Journal of Economic History*, III, Supplement (1943), 66-77.

Hazard, Erskine, "History of the Introduction of Coal into Philadelphia," *Memoirs of the Historical Society of Pennsylvania*, II (1828).

Heindel, Ned D., "The Centennial of the First Class Trip on Pennsylvania Canals: The Voyage of the Molly-Polly-Chunker," *Pennsylvania Folklife*, XXXVI(2), (1986-1987), 79-89.

PENNSYLVANIA (Cont.)

Henry, Matthew S., *History of the Lehigh Valley* (Easton, Pa: Bixler & Corwin, 1860).

Hensel, William U., "An Early Canal Project," *Lancaster Historical Society Papers*, XVII ((1913), 101-105.

Hesser, A. A., "The Schuylkill Canal," *Schuylkill County Historical Society Publications*, IV (1913), 9-17.

"The Heyday of the Schuylkill Navigation Company," *Historical Review of Berks County*, IV (1938), 114.

Heydinger, Earl J., "American Canal Tunnels," *CC*, No. 39 (Summer 1977), 9.
————————, "Banker Jay Cooke, Former Packet Runner," *CC*, No. 58 (Spring 1982), 7.
————————, "Boar's Nest: Susdquehanna Packet ~Smoker' for Homebound Arkers and Raftmen," *CC*, No. 36 (Autumn 1976), 5.
————————, "Canal Around A Bridge," *CC*, No. 47 (Summer 1979), 6.
————————, "Canalling North-South Across Pennsylvania," *CC* No. 46 (Spring 1979), 10-11.
————————, "Columbia: Leading Canal Port," *CC*, No. 48 (Autumn 1979), 11-13.
————————, "Completion Costs of Erie [Extension] Canal, 1845," *CC*, No. 68 (Autumn 1984), 16.
————————, "Conewago Canal, Pennsylvania;s First," *CC*, No. 59 (Summer 1982), 9-10.
————————, "Conestoga Navigation, 1829-1965," *CC*, No. 41 (Winter 1978), 6, 10.
————————, "The Decline of the Section Boats," *CC*, No.45 (Winter 1979), 4-5.
————————, "The Delaware and Schuylkill Canal," *CC*, No. 54 (Spring 1981), 4-5, 15, [reprint from *Bulletin of the Historical Society of Montgomery County*, Spring 1979, XXI (4), 358-363.
————————, "Early French Creek Traffic," *CC*, No. 47 (Summer 1979), 14.
————————, 1814 Coal Ark Departure from Mauch Chunk, *CC*, No.58 (Spring 1982), 3.
————————, "The 1833 Warren Convention and Transshipment," *CC*, No. 38 (Spring 1977), 10.
————————, "The 1828 Coal Canal to Pine Grove," *CC*, No. 55 (Summer 1981), 6-7.
————————, "The Erie Extension in Erie County," *CC*, No. 34 (Spring 1976), 7.
————————, "Experience Killed the North Branch," *CC*, No. 37 (Winter 1977), 7.
————————, "First Tunnel in the U.S.A.," *CC*, No. 5 (Summer 1968), 3.

PENNSYLVANIA Cont.)

Heydinger, Earl J., "The Four-Mile Canal Tunnel," *CC*, "No. 38 (Spring 1977), 8.
——————————, "The Franklin Line of the Erie Extension," *CC*, No. 37 (Winter 1977), 3-4, 12
——————————, "Grant's Hill Revisited," *CC*, No. 47 (Summer 1979), 16.
——————————, "Hopkins Canal at Conewago Falls," *CC*, No. 57 (Winter, 1982), 4-5.
——————————, "Jay Cooke and the 1858 Canal Sales," *CC*, No. 67 (Summer 1984), 6, 13.
——————————, "Josiah White and His Bear Trap Navigation," *CC*, No. 41 (1978), 8.
——————————, "Josiah White, Improvements on the Schuykill, 1813-18," *CC*, No. 51 (Summer 1980), 10-11.
——————————, "Kernsville Dam on the Schuylkill Canal," *Historical Review of Berks County*, XV (1949), 145-147.
——————————, "Mainline Tunnel of 1829," *CC*, No. 37 (Winter 1977), 11.
——————————, "The North Branch and its Chemung Connection," *CC*, No.58 (Spring 1982), 5
——————————, "Notes on the Pennsylvania and Ohio Canal," *CC*, No. 57 (Winter 1982), 16.
——————————, "Opening of the P[ennsylvania] & O[hio] Canal," *CC*, No. 67 (Summer 1984), 14.
——————————, "Packet Boat Accident!! Loss of Life!!" *CC*, No. 41 (Winter 1978)
——————————, "Packets, 'Huntingdon To Philadelphia'," *CC*, No. 65 (Winter 1984), 7.
——————————, "The Pennsylvania and Ohio Canal," *CC*, No. 35 (Summer 1976), 10-13 [reprinted from *Towpaths*, publication of the Ohio Canal Society].
——————————, "Penna. Canal Machine and Corruption," *CC*, No. 65 (Winter 1984). 10.
——————————, " Plane Power: Chain, Iron Band, Rope," *CC*, No. 35 (Summer 1976), 1, 9.
——————————, "The Plunder Act-A Boon to the P & O," *CC*, No. 60 (Autumn 1982), 11, 12.
——————————, "Running for Canal Boats," *CC*, No. 52 (Autumn 1980), 12.
——————————, "Schuylkill Canal Docks," *CC*, No. 69 (Winter 1985), 7.
——————————, "The Schuylkill Canal through Berks County," *Historical Review of Berks County*, XXXVII (1972), 128-132.
——————————, "Schuylkill Navigation in Reading, 1832," *CC*, No. 50 (Spring 1980), 11-12.
——————————, "Susquehanna Canal Chartered in 1783," *CC*, No. 43 (Summer 1978), 4-5, 7.

PENNSYLVANIA (Cont.)

Heydinger, Earl J., "Susquehanna Coal Ports," *CC*, No. 40 (Autumn 1977), 9-10.

——————————————, "Three Allegheny Portage Railroads," *CC*, No 26 (Fall 1973), 1, 8.

——————————————, "Walnut Street Aqueduct," *CC*, No. 40 (Autumn 1977), 5.

——————————————, "Washington Visits Penna. Canals," *CC*, No. 69 (Winter 1985), 9.

——————————————. "Watch out, Erie Canal, The Section Boats are Coming!" *CC*, No. 63 (Autumn 1983), 5-7.

A History of the Monongahela Navigation Company by an Original Stockholder (Pittsburgh: Bakewell & Marthens, 1873, reprinted by Pennsylvania Canal Society, 1978).

Hobbs, Gibson, (ed.), "Notes of a Trip to the Union Canal & Up the Schuylkill-1826," by John H. Cocke, Jr., *CC*, No. 53 (Winter, 1981), 11-15.

Hoffman, Daniel B," History Corrected, Heydinger Satisfied," *CC*, No. 53 (Winter 1981), 3; see also Heydinger, Earl J, "Coaldirt-The Last Word," *CC*, No. 53 (1981), 3, 15.

Hoffman, John N., *Anthracite in the Lehigh Region of Pennsylvania* (Washington D.C: Gov't Printing Office, 1968).

Hoffman, Philip J. "Johnstown Canal Basin," *CC*, No. 6 (Fall 1968), 3.

Hone, Philip, *Diary of Philip Hone*, 2 vols., Alan Nevins (ed.), (New York: Dodd, Mead & Co., 1927).

Houtz, Harry, "Abner Lacock, Beaver County's Exponent of the American System," *WPHM*, XXII (1939), 177-187.

Ilisevoch, Robert D., "Huidekoper—Canal Promoter," *CC*, No. 79 (Summer 1987), 9-10, 16.

Ilisevich, Robert D and Carl K Burkett, "The Canal Through Pittsburgh: Its Development and Physical Character," *WPHM*, LXVIII (1985), 351-372.

Ingham, S. D., (compiler), *Canal and Railroad Laws of Pennsylvania prepared and published under authority of a resolution of the House of Representatives, passed the sixteenth day of June, A. D. 1836.* (Harrisburg: Theo. Penn—Printer, 1836)

PENNSYLVANIA (cont.)

Jacobs Harry A., *The Juniata Canal and Old Portage Railroad* (Hollidaysburg, Pa: Blair County Historical Society, 1941, 1969).
————————, "The Juniata Canal and Old Portage Railroad," *PDIA*, XII, #3, (1944), 3-14; XII, # 4, 3-11; XII, # 5, 3-8, 30.

James, Bessie, *Ann Royall's America* (New Brunswick, NJ: Rutgers University Press, 1972).

"A Jaunt to Reading in 1825," *The Historical Review of Berks County*, III, No. 2 (Jan. 1938), 41-43.

Jenkins, Howard M. (ed.), *Pennsylvania: Colonial and Federal*, 3 vols. (Philadelphia: Pennsylvania Historical Publishing Association, 1903), Vol. III, Chapter VI, "Internal Improvements", 254-334.

John, J.J., "The Story About One of the First Chartered Railroads in America," *Publications of the Historical Society of Schuylkill County*, I (1907), 343-411.

Johnson, George B., Earl B. Giles and Ralph Michaels, "Johnstown and the Pennsylvania Canal," *CC*, No. 63 (Summer 1983), 1-14.

Johnson, George B., "Adolph Von Graffius and the Main Line in 1841," *CC* No. 50 (Spring 1980), 13-15.
————————, "Ann Royall Visits Saltsburg," *CC*, No 47 (Summer 1979), 8-9.
————————, "The Canal Store in Centreville," *CC*, No. 66 (Spring, 1984), 3-12.
————————, "The Last of a Breed: Canal Boatmen's Association Meetings in or Near Saltsburg," *CC*, No. 56 (Autumn 1981), 11-15.
————————, "More Shipments of Salt," *CC*, No. 62 (Spring 1983), 12-15.
————————, "Murder on the Canal!" *CC*, No. 59 (Summer 1982), 10-15.
————————, "Mysteries of the Canal Boat Ledger," *CC* No. 71 (Summer 1985), 1-12.
————————, *Saltsburg and the Pennsylvania Canal* (Saltsburg: Pa, author, 1984).
————————, "Saltsburg and the Pennsylvania Canal," *CC*, No. 46 (Spring 1979), 1, 12-15.
————————, "The Saltsburg Borough Council Minutes and the Pennsylvania Canal, 1850-1866," *CC*, No.48 (Autumn 1979), 4-7.
————————, "Saltsburg Canal Diary," *CC*, No. 51 (Summer 1980), 3-7.

PENNSYLVANIA (Cont.).

Johnson, George B., "Saltsburg Canal People," *CC*, No. 54 (Spring 1981), 10-15.
————————————, "Samuel S. Jamison, Canal Contractor, Supervisor, Senator," *CC*, No. 68 (Autumn 1984), 3-11.
————————————, "A Shipment of Salt," *CC*, No. 58 (Spring 1982), 1-2.
————————————, "The Stone Carver--A Happy Canal Incident," *CC*, No. 64 (Autumn 1983), 8-9.
————————————, "Whiskey Shipped on the Western Division," *CC*, No. 67 (Summer 1984), 7.

Johnson, Leland R., *The Davis Lock and Dam: 1870-1922* and accompaning *Portfolio* (12 plates) (Pittsburg, Pa: U. S. Army Corps of Engineers, 1985).
————————————, *The Headwaters District: A History of the Pittsburgh District, U.S. Army Corps of Engineers* (Washington, D.C: Government Printing Office, [1979]).

Johnston, John Willard, *Reminiscences and Descriptive Account of the Delaware Valley and its Connections Aiming to extend from Pond Eddy to Narrowsburg*, 2 Mss volumes (Narrowsburg, N.Y:, 1900, reprint, Town of Highland [N.Y.] Cultural Resources Commission, 1987).

Jones, Chester L., *The Economic History of the Anthracite-Tidewater Canals* (Philadelphia: John C,. Winston, 1908).
————————————, "The Anthracite-Tidewater Canals," *The Annals of the American Academy of Political and Social Science*, XXXI (January-June 1908), 102-116.

Jones, Samuel, *Pittsburgh in the Year 1826* (Pittsburgh: Johnson and Stockton, 1826, reprinted, Arno & New York Times, 1970).

Jordan, William M., "Canals of the Lebanon Valley," *Pennsylvania Geology*, XIV (April 1983), 2-5.

"Journal of a Voyage on the Raging Canal," *WPHM*, LXIII (1980), 273-276.

Kerstetter, Norman C., "A Woman's Life on a Canalboat," *CC*, No. 31 (Winter 1975), 12.

Kirby, Richard S., "William Weston and His Contributions to Early American Engineering," *Newcomen Society Transactions* [English], XVI (1935-1936), 111-127.

PENNSYLVANIA (Cont.)

Kizen, Ben "Canaller Recalls Boating Years," *CC*, No. 54 (Spring 1981), 7 [from *The Easton Express*, April 1, 1949].

Klein, Theodore B., *The Canals of Pennsylvania and the System of Internal Improvement*, (Wm. Stanley Ray, State Printer of Pennsylvania, 1901, reprint, Canal Press, Inc., 1973).

Knies, Michael, "Industry, Enterprise, Wealth and Taste: The History of Mauch Chunk, 1791-1831," *CCHT*, IV (1985), 17-44.
----------------, "Stone Coal in the Switzerland of America," *Pennsylvania Heritage*, XV (Winter 1989), 10-17.

Koehler, John (contributor), "The Beaver Meadow Railroad," based on *Annals of the Sugarloaf Association*, II (1935), Hazleton, Pennsylvania, *CC*, No. 72 (Fall 1985), 9-12.

Kussart, Sarepta, *Navigation on the Monongahela River*, 2 vols, (Pittsburgh: Daily Republican, 1929).
----------------, *The Allegheny River* (Pittsburgh: Burgum Printing Co., 1938).

Landes, D.B., "The Conestoga River At Lancaster," *Journal of the Lancaster County Historical Society*, XXXVI (1932), 261-71.

Larson, Henrietta M., "Jay Cooke's Early Work in Transportation," *PMHB*, LIX (1935), 362-375.
----------------, *Jay Cooke, Private Banker* (Cambridge, Mass: Harvard Univ. Press, 1936).

"Lawsuit Eyes Life on Canals," *CC*, Nos. 76-77 (Fall-Winter 1986-1987) 15-20, 24.

"Lawsuit Follows Drowning of Mule in Canal," *CC*, No. 79 (Summer 1987), 15, 16.

LeRoy, Edward, *The Delaware and Hudson Canal: A History* (Honesdale: Wayne County Historical Society, 1950).
----------------, *Delaware & Hudson Canal and its Gravity Railroads* (Honesdale, Pa: Wayne County Historical Society, 1980).
----------------, "The Delaware and Hudson Canal," *CC*, No. 61 (Winter 1983), 15-16.
----------------, "Delaware and Hudson Canal Pioneer Coal Carrier," *PDIA*, XIII, #11 (1945), 3-9; XIII, # 12 , 3-7; XIV, # 1 (1946), 3-8; XIV, # 2&3, 9-12; XIV, # 4, 28-31; XIV. # 5, 6-10; XIV, # 6, 8-14; XIV, # 7, 23-26; XIV, # 8, 18-22; XIV, # 10 (1946)4-8.

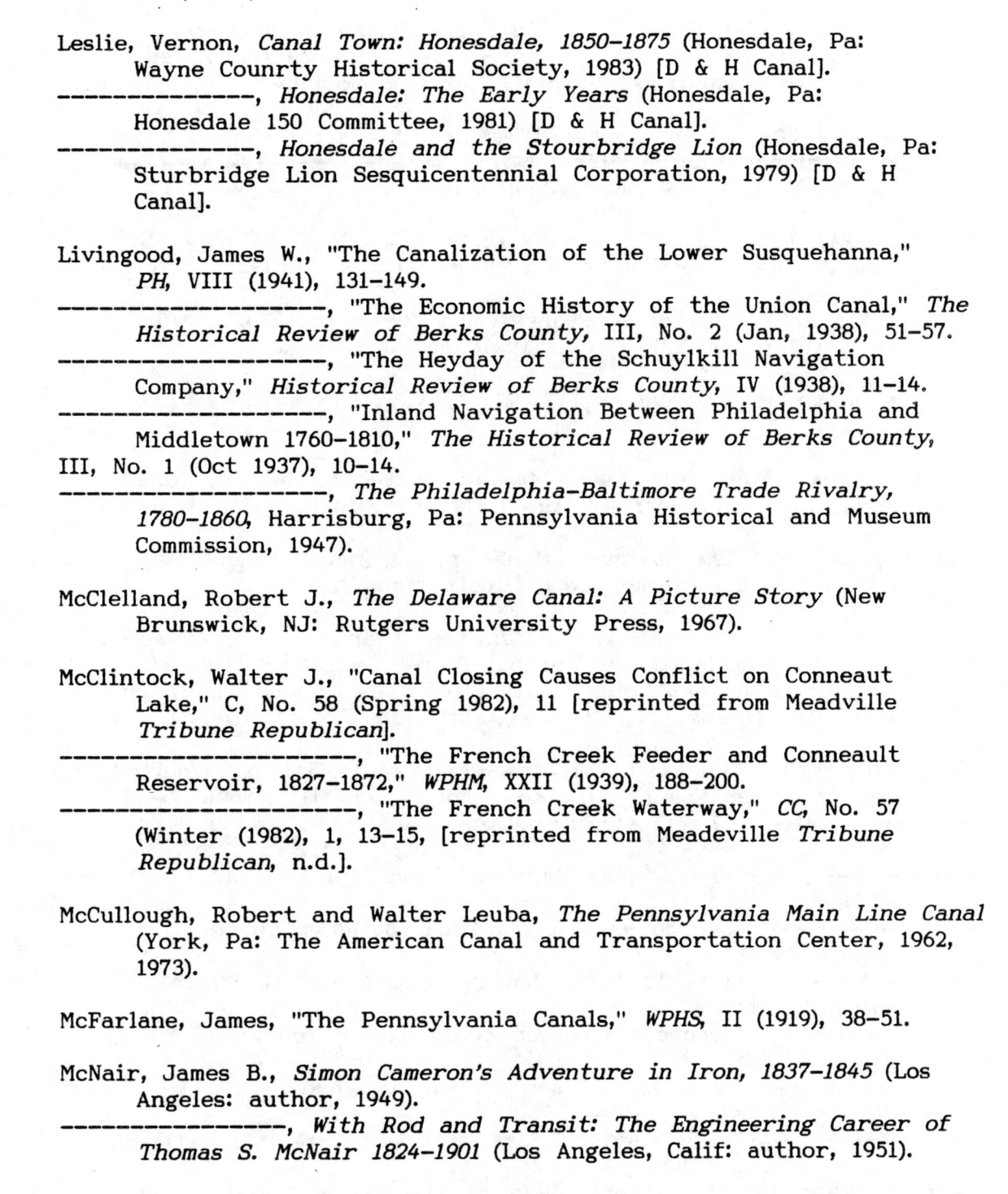

PENNSYLVANIA (Cont.)

Leslie, Vernon, *Canal Town: Honesdale, 1850-1875* (Honesdale, Pa: Wayne Counrty Historical Society, 1983) [D & H Canal].
——————, *Honesdale: The Early Years* (Honesdale, Pa: Honesdale 150 Committee, 1981) [D & H Canal].
——————, *Honesdale and the Stourbridge Lion* (Honesdale, Pa: Sturbridge Lion Sesquicentennial Corporation, 1979) [D & H Canal].

Livingood, James W., "The Canalization of the Lower Susquehanna," *PH*, VIII (1941), 131-149.
——————, "The Economic History of the Union Canal," *The Historical Review of Berks County*, III, No. 2 (Jan, 1938), 51-57.
——————, "The Heyday of the Schuylkill Navigation Company," *Historical Review of Berks County*, IV (1938), 11-14.
——————, "Inland Navigation Between Philadelphia and Middletown 1760-1810," *The Historical Review of Berks County*, III, No. 1 (Oct 1937), 10-14.
——————, *The Philadelphia-Baltimore Trade Rivalry, 1780-1860*, Harrisburg, Pa: Pennsylvania Historical and Museum Commission, 1947).

McClelland, Robert J., *The Delaware Canal: A Picture Story* (New Brunswick, NJ: Rutgers University Press, 1967).

McClintock, Walter J., "Canal Closing Causes Conflict on Conneaut Lake," *C*, No. 58 (Spring 1982), 11 [reprinted from Meadville *Tribune Republican*].
——————, "The French Creek Feeder and Conneault Reservoir, 1827-1872," *WPHM*, XXII (1939), 188-200.
——————, "The French Creek Waterway," *CC*, No. 57 (Winter (1982), 1, 13-15, [reprinted from Meadeville *Tribune Republican*, n.d.].

McCullough, Robert and Walter Leuba, *The Pennsylvania Main Line Canal* (York, Pa: The American Canal and Transportation Center, 1962, 1973).

McFarlane, James, "The Pennsylvania Canals," *WPHS*, II (1919), 38-51.

McNair, James B., *Simon Cameron's Adventure in Iron, 1837-1845* (Los Angeles: author, 1949).
——————, *With Rod and Transit: The Engineering Career of Thomas S. McNair 1824-1901* (Los Angeles, Calif: author, 1951).

PENNSYLVANIA (Cont.).

MacReynolds, George, *Place Names in Bucks County Pennsylvania Alphabetically arranged in an Historical Narrative*, 2d ed (Doylestown, Pa: Bucks County Historical Society, 1942, 1976).

Madden, J. Hayward (contributor), "Building the Union Canal, A Major Engineering Triumph," [from *Niles Register*, XXI, 132, October 28, 1826], *CC*, No. 9 (Summer 1969), 6.

Marchwinski, Patricia J., "Following History Through the Erie Extension Canal," *The Journal of Erie Studies*, XII. No. 2 (1983), 1-17.

Martin, Asa E., "Lotteries in Pennsylvania Prior to 1833," *PMHB*, XLVII (1923), 307-327, XLVIII (1924), 66-93, 159-180.

Martineau, Harriet, "Travels of . . ." *CC*, No. 52 (Autumn 1980), 11-12.

"Mauch Chunk," from *Picturesque America* edited by William Cullen Bryant, Vol. I; reprinted in *CC*, No. 69 (Winter 1985), 1, 3-5.

Mauch Chunk, Pa., The Switzerland of America (Mauch Chunk, Published by Tosh's Dept. Store, n.d.), illustrations.

Maximilian, Prince of Wied, *Travels in the Interior of North America* [Part I], translated by Hannibal Evans Lloyd. (London: 1843, reprint in Reuben Gold Thwaites (ed.), *Early Western Travels, 1748-1846*, volume XXII, 1905, reprint, AMS Press, 1966).

Mayo, Robert S., "The Fifth Columbia-Wrightsville Bridge," *Journal of the Lancaster County Historical Society*, LXXIII (1969), 26-33.
--------------, "The First Susquehanna Canal," *AC*, No. 21 (May 1977), 6, 7.
--------------, "Franklin on Depth of Canals," *CC*, No. 36 (Autumn 1976), 4.
--------------, "George Washington: Canal Builder," *AC*, No. 18 (August 1976), 6.
--------------, "Improving the Lower Susquehanna River: 1801," *CC*, No. 51 (Summer 1980), 8-9.
--------------, "Second Honeymoon on Allegheny Tour," *CC*, No. 59 (Summer 1982), 5-6.
--------------, "'Stopper Hitches' on the Allegheny Portage Railroad," *AC*, No.11 (November 1974), 3.
--------------, "Water-Saving Locks," *CC*, No. 37 (Winter 1977), 10.

Metheny, Thomas Bannon, "Early Days on the Beaver and Erie," *CC*, No. 68 (Autumn 1984), 14.
--------------------, "Life on the Beaver and Erie Canal," ed. by Denver L. Walton, *CC*, No. 19 (Winter 1972), 3-4.

PENNSYLVANIA (Cont.)

Metz, Lance E., "Josiah White's Zoo," *CC*, No. 60 (Autumn 1982, 11, 12.

----------------, (Exhibit Curator), *Robert H. Sayre, Engineer, Entrepreneur, Humanist, 1824-1917*, (Easton, PA: Hugh Moore Historical Park and Museum, 1985).

----------------, "'Rome Haul' Movie Filmed on Lehigh Canal," *CC*, No. 51 (Summer 1980), 13.

----------------, (researcher), *Cap't Sherman's Guide to Hugh Moore Park* (Easton, Penna: Center for Canal History and Technology, 1988).

Michaels, Ralph, "Portage Railroad Accidents," C, No. 67 (Summer 1984), 1-3.

Miller, John P., *The Lehigh Canal, A Thumb Nail History, 1829-1931*, (Printed for the Sesquicentennial of the opening of the Lehigh Canal, 1979).

Mix, Richard L., "Trips and Travelers on the West Branch Canal," *CC*, No.44 (Autumn 1978), 6-8.

Morell, William H, *Report Relative to the Condition of the North Branch Canal* (Philadelphia: T.K. & P.O. Collins, 1856).

Montz, W. Curtis, "Water Transportation on the Susquehanna," *Proceedings and Collections of the Wyoming Historical and Geological Society*, XXII (1970), 20-39.

Morris, Ellwood, "Description of the Bear Trap Sluice Gates of the Lehigh Descending Navigation," *Journal of the Franklin Institute*, XXXII (Dec 1841), 361-368 [illustrated].

Morton, Eleanor, pseud. (Elizabeth Gertrude Stern), *Josiah White* (New York: Stephen Daye Press, 1946.

Murphy, Julius W., "The Canal Comes to Pittsburgh," *CC*, No. 41 (1978), 7, 10.

Murray, Elsie, *The North Branch Canal* (Athens, Pa: 1941).
--------------, *Stephen C. Foster at Athens: His First Composition* (Athens, PA: Tioga Point Museum, 1941).

Murray, Louise Welles, *A History of Old Tioga Point and Early Athens Pennsylvania* (Athens, Pa: author, 1908), Chapter XX, "Development of Highways and Transportation," 519-536.

PENNSYLVANIA (Cont.)

Myers. Richmond E., "Flatboat on the River [Susquehanna]," *PDIA*, XXII, # 1 (1953), 3-10, 23-28.

————————————, "The Long Crooked River and its Human Utilization [Susquehanna]," *PDIA*, XX, # 2 (1952), 12-17.

————————————. "The Story of Transportation on the Susquehanna River," *New York History*, XXIX (1948), 157-169.

"The Newark Camera Club's Best," *CC*, No. 30 (Fall 1974), 1-2, 7-8; No. 31 (Winter 1975), 8.

Newell, James D., "The Beaver Division Canal in New Castle," [from *New Castle News*, January 1, 1923], *CC*, No. 52 (Autumn 1980), 5-6.

"New Picture of Historic Old Lemon House," *PDIA*, XVI, # 10 (1948), 20-1.

"A New Yorker Reports on Pennsylvania's Main Line," [from *Hazard's Register*, XVI (August 1835)], *CC*, No. 57 (Winter 1982), 11-15.

The North American Tourist, (New York: A.T. Goodrich, n.d.).

"Notice of the Sandy and Beaver and Mahoning Canal," *Journal of the Franklin Institute*, XIX (1835), 297-302; "Report of the President. . .of. . .," XXI (1836), 23-27.

Oberholtzer, Ellis P., *Jay Cooke, Financier of the Civil War*, 2 vols (Philadelphia: George W. Jacobs & Co., 1907) I, 40-50.

"Observations on the Htdraulic Cement, used on the Pennsylvania Canal," *Journal of the Franklin Institute*, II (1826), 287-288.

Oliphant, J. Orin, "How Lewisburg Became a Canal Port," *Proceedings of the Northumberland County Historical Society*, XXI (1957), 37-66.

————————————, "A Journey from Philadelphia to Lewisburg 108 Years Ago," *PDIA*, XXIII, # 1 (1954), 20-23.

Osborne, Peter, III., "The Delaware and Hudson Canal Company's Enlargement and the Roebling Connection," *CCHT*, III (1984), 119-134.

Parton, W. Julian, *The Death of a Great Company* (Easton, Pa: Center for Canal History and Technology, 1986).

Patschke, M. Luther, "The Union Canal. . .Yesterday and Today: Boyhood Memories of the Old Canal," *CC*, No 56 (Autumn 1981), 1, 6-9.

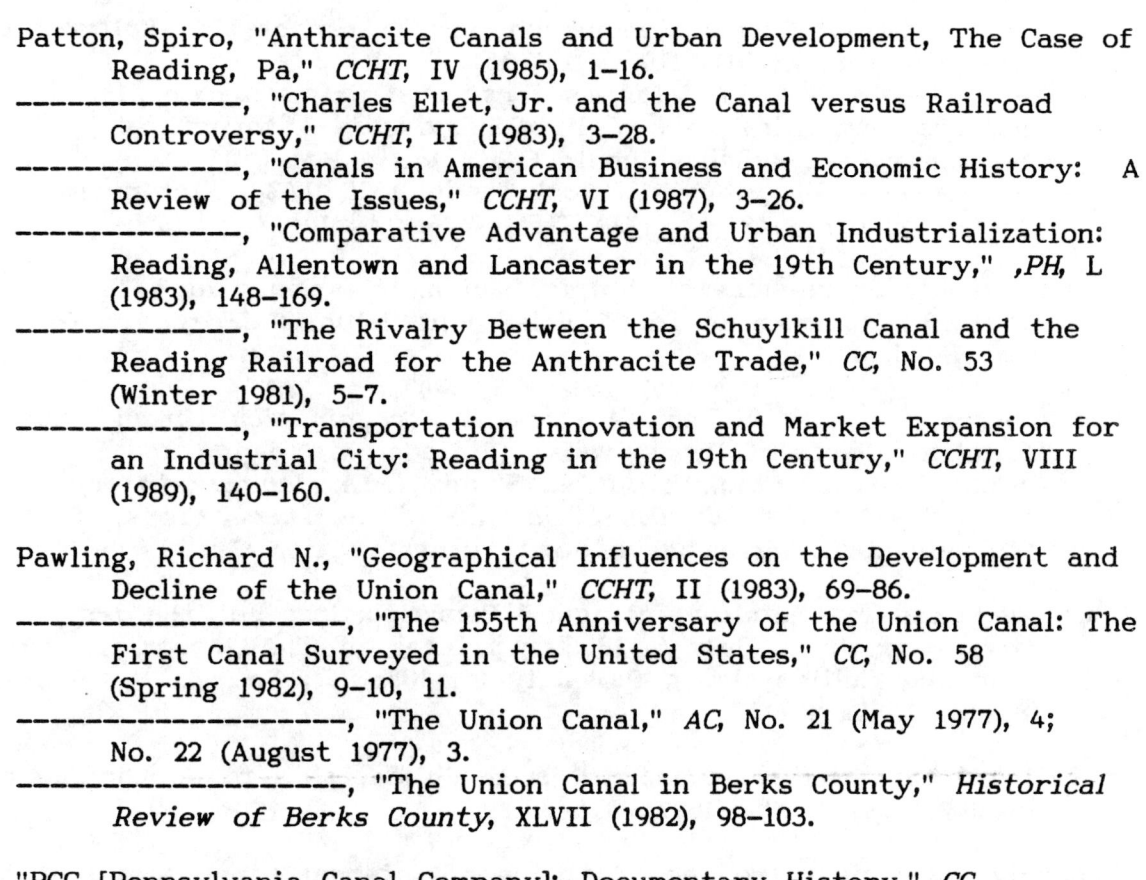

PENNSYLVANIA (Cont.).

    Patton, Spiro, "Anthracite Canals and Urban Development, The Case of Reading, Pa," *CCHT*, IV (1985), 1-16.

    ---------------, "Charles Ellet, Jr. and the Canal versus Railroad Controversy," *CCHT*, II (1983), 3-28.

    ---------------, "Canals in American Business and Economic History: A Review of the Issues," *CCHT*, VI (1987), 3-26.

    ---------------, "Comparative Advantage and Urban Industrialization: Reading, Allentown and Lancaster in the 19th Century," *,PH*, L (1983), 148-169.

    ---------------, "The Rivalry Between the Schuylkill Canal and the Reading Railroad for the Anthracite Trade," *CC*, No. 53 (Winter 1981), 5-7.

    ---------------, "Transportation Innovation and Market Expansion for an Industrial City: Reading in the 19th Century," *CCHT*, VIII (1989), 140-160.

    Pawling, Richard N., "Geographical Influences on the Development and Decline of the Union Canal," *CCHT*, II (1983), 69-86.

    --------------------, "The 155th Anniversary of the Union Canal: The First Canal Surveyed in the United States," *CC*, No. 58 (Spring 1982), 9-10, 11.

    --------------------, "The Union Canal," *AC*, No. 21 (May 1977), 4; No. 22 (August 1977), 3.

    --------------------, "The Union Canal in Berks County," *Historical Review of Berks County*, XLVII (1982), 98-103.

"PCC [Pennsylvania Canal Company]: Documentary History," *CC*, No. 73 (1986), 5-11.

"Pennsylvania Canal Trade: Arrivals and Departures from April 6 to April 22," [from *Harrisburg Chronicle*, Monday 22, 1833] *CC*, No 40 (Autumn 1977), 11.

Petrillo, F. Charles, *Anthracite and Slackwater: The North Branch Canal, 1828-1901*, Easton Pennsylvania: Center for Canal History and Technology, 1986).

    --------------------, "The Pennsylvania Canal Company (1857-1926): The New Main Line Canal, Nanticoke to Columbia," *CCHT*, VI (1987), 83-112.

Pharo, Elizabeth B. (ed.), *Reminiscences of William Hasell Wilson (1811-1902)*, (Philadelphia: Privately printed, 1937)

PENNSYLVANIA (Cont.)

Powell, H. Benjamin, "Coal and Pennsylvania's Transportation Policy, 1825-1828," *PH*, XXXVIII (1971), 134-151.

—————————————, *Philadelphia's First Fuel Crisis: Jacob Cist and the Developing Market for Pennsylvania Anthracite* (University Park: Pennsylvania State Univ. Press, 1978).

—————————————, "Schuylkill Coal Trade, 1825-1842," *Historical Review of Berks County*, XXXVIII (1972-1973), 14-17.

Power, Tyrone, *Impressions of America during the years 1833, 1834, and 1835*, 2d American Ed., 2 vols, (Philadelphia: Carey, Lea & Blanchard, 1836).

Prolix, Peregrine [P.H. Nicken], *A Pleasant Peregrination through the Prettiest Parts of Pennsylvania*, (Philadelphia: Grigg and Elliot, 1836); also reprinted, William H. Shank, (ed.), *Journey through Pennsylvania--1835, By Canal, Rail and Stage Coach*, (York, Pennsylvania: American Canal and Transportation Center, 1975).

"PRR Denounced for Mainline Intrigue," Correspondence of the *Daily News*, Harrisburg, July 4, 1855, reprinted in *The Democratic Standard*, Hollidaysburg, Pa., July 25, 1855, *CC*, No. 58, (Spring 1982), 12-15.

"A Railroad and Canal-Boat Journey from Philadelphia to Northumberland in 1835," *PMHB*, XXI (1897), 414-5.

"Pennsylvania Society fro the promotion of Internal Improvements in the Commonwealth," *Journal of the Franklin Institute*, I (1826), 10-16, 64, 71-77, 134-138, 197-200; includes frports, and the instructions to and the reports from William Strickland.

Pennsylvania Writers Project [WPA], *Pennsylvania Cavalcade* (cosponsered by the Pennsylvania Federation of Historical Societies; Philadelphia: Univ. of Pennsylvania Press, 1942); "The Schuylkill Canal," pps. 327-337; "The Allegheny Portage Railroad," pps. 385-393; "The Erie Extension Canal," pps. 394-407.

"Railroads of Pennsylvania in 1848," [from *Appleton's Railroad and Steamboat Companion*], *CC*, No. 71 (Summer 1985), 14-16.

Rapp, R. Francis, "Lehigh and Delaware Division Canal Notes," *Bucks County Historical Society Papers*, IV (1917), 600-606.

PENNSYLVANIA (cont.)

"Rare Photograph Along Old Portage Railroad," *PDIA*, XIII, # 1 (1945), 10-11.

Reck, Bruce and David, "Tracing the Erie Extension Canal Through Erie County," *CC*, No. 50 (Spring 1980), 9-11.

Reed, John E., "The Erie Extension of the Pennsylvania Canal," *Erie County Annals*, 1952, 5-8.

Reed, John E. and John I, Cretzinger, "The Old Erie County Canal, Its History and Service [Titles sometimes vary]," *PDIA*, XVI, Nos. 8 and 9 (1948), 3-7: XVI, # 10 [sic. September], 3-7; XVI # 11 [sic. October], 14-19; XVI, # 12 (Dec. 1948), 8-10; XVII, # 1 (1948), 10-13; XVII, # 2 (1949), 3-9; XVII, # 3, 18-22; XVII, # 4, 15-18; XVII, # 5, 19-20; XVII, # 6, 24-32.

Reid, R.D., [map] "The Waterways of Eastern Ohio and Western Pennsylvania, Past, Present and Future, 1983," *CC*, No. 71 (Summer 1985), 8-9.

Reiser, Catherine E., *Pittsburgh's Commercial Development, 1800-1830* (Harrisburg, Pa: Pennsylvania Historical and Museum Commission, 1951).

"Report of a Committee of the Stockholders of the Conestoga Navigation Company, made July 1, 1822," *HR*, X (1832), 54-9.

"Report of Moncure Robinson, Principal Engineer upon the Allegheny Portage" (21 November 1829), *Pennsylvania House Journal*, 1827-1828, II, Doc. 138, reprinted in *Transactions of the American Society of Civil Engineers*, XV (1886), 183-202.

"Restoration of Canals for Recreational Purposes," *PDIA*, XIII, # 9 (1945), 10-11.

Revinius, Willis M., *A Wayfarer's Guide to the Delaware Canal between Bristol and Easton* (Doylestown, Pa: Bucks County Park Foundation, 1984), several revisions.

Reynolds, John E., *In French Creek Valley* (Meadville, Pa: The Crawford County Historical Society, 1938).

Rhoads, Willard R., "The Pennsylvania Canal," *WPHM*, XLIII (1960), 203-238, [reprinted in *CC*, No. 32/33 (Winter 1976), 1-14].
———————————— "Pennsylvania's North Branch Canal," (with supplement: Living and Working on the North Branch Canal: Reminiscences and Aspects by Edwin M. Barton), *The Columbian*, II (October 1962), pps. ???

PENNSYLVANIA Cont.).

Roberts, Solomon W., "Reminiscences of the First Railroad over the Allegheny Mountains," *PMHB*, II (1878), 370-393.

Roberts, F. P., "The Monongahela River," *CC*, No. 75 (Summer 1986), 5-16.

"Roebling Describes Aqueduct," *CC*, No. 78 (Spring 1987), 10, 16.

Rosenberger, Homer T., *The Philadelphia and Erie Railroad, Its Place in American Economic History* (Potomac, Maryland: The Fox Hills Press, 1975).
————————————, "Philadelphia's Influence On Pennsylvania Transportation Development," *PDIA*, XVII, # 12 (1949), 3-11.

Royall, Anne, "Journey to Mauch Chunk," *CC*, No. 35 (Summer 1976), 15, 16 [extracted from *Pennsylvania, or Travels Continued in the United States* 2 vols, 1829].

Rubin, Julius, "Canal or Railroad? Imitation and Innovation in Response to the Erie Canal in Philadelphia, Baltimore and Boston," *Transactions of the American Philosophical Society*, New Series, CI, pt. 7, 1961.

Rung, Albert M., "Canaling was a Thrilling, Rugged Life," *CC*, No. 4 (Spring 1968), 5.
————————————, "Canal Travel Led to Romance," *CC*, No. 14 (Fall 1970), 1.
————————————, "Did Whiskey Build a Canal?" *CC*, No. 5 (Summer 1968), 4.
————————————, "Life on the Juniata Division Recalled," *CC*, No. 10 (Fall 1969), 3, 4.
————————————, "The Packet 'Dr. William Lehman' Leaves Huntingdon, 1832," *CC*, No.3 (Winter 1968), , 3.
————————————, "The Strange Story of George von Breck," *CC*, No. 13 (Summer 1970), 3-4.
————————————, "William Lehman 'The Dreamer'," *CC*, No.44 Autumn 1978), 4-5 [reprinted from *Rung's Chronicles of Pennsylvania History* (1977)].

Rupp, Isreal Daniel, *History of Northampton, Lehigh, Monroe, Carbon and Schuylkill Counties* (Harrisburg: Hickok and Cantine, 1845).

"J.M.S.","General Abner Lacock," *PMHB*, IV (1880), 202-8

Sanderlin, Walter S., "The Expanding Horizons of the Schuylkill Navigation Company, 1815-1870," *PA*, XXXVI (1969), 174-191.

PENNSYLVANIA (Cont.).

Sanderson, Dorothy H, *The Delaware and Hudson Canalway: Carrying Coals to Roundout*, 2d edition (Ellensville, N.Y: Roundout Valley, 1965, 1974).

Sappee, Nathan D., "Spoilation and Encroachment in the Conemaugh Valley Before the Johnstown Flood of 1889," *WPHM*, XXIII (1940), 25-48.

Sayenga, Donald (contributor), "Allegheny Aqueduct Removed," *CC*, No. 79 (Summer 1987), 12.
——————————, "America's first Wire Rope Factory," *CC*, No. 55 (Summer 1981), 1, 3-5.
——————————, *Ellet & Roebling, The Amazing Tale of Friendship and Rivalry Between Two of America's Greatest Engineers* (York, Pa: American Canal and Transportation Center, 1983).
—————————— (ed.), "John Augustus Roebling and the Public Works of Pennsylvania by Dr. Hubertis M. Cummings," *CCHT*, III (1984), 119-135.
——————————, "Roebling's Pittsburgh Aqueduct," *CC*, No. 61 (Winter 1983), 3-6.
——————————, "The Original Pittsburgh Aqueduct," *CC*, No. 59 (1982), 1-4.
——————————, "T. Haskins Dupuy's Survey of the Pennsylvania Main Line Canal," *CCHT*, VI (1987), 71-82.
——————————, "The Untried Business: An Appreciation of White and Hazard," *CCHT*, II (1983), 105-128.

Schuleen, Ernest T., "Two Rivers and a Village, the Story of Safe Harbor," *Journal of the Lancaster County Historical Society*, LXXXV (1981), 82-125.

Schnure, William M., "Boat Building in Selinsgrove," *Snyder County Historical Society*, II(6) (1944), 3-6.
——————————————, "The Pennsylvania Canal Along the Susquehanna," *Proceedings of the Northumberland County Historical Society*, XVI (1948), 89-101.
——————————————, "The Ragin Canal," *Now and Then*, IX (1949), 86-91.

Schotter, H. W., *The Growth and Development of the Pennsylvania Railroad Company, 1846 to 1926* (Philadelphia: Allen, Lane and Scott, 1927).

Scriven, George B., "The Susquehanna and Tidewater Canal," *Maryland Historical Magazine*, LXXI (1976), 522-526.

PENNSYLVANIA (Cont.)

"A Second Pennsylvania Canal," from *Indiana Progress*, May 1, 1879, *CC*, No 44 (Autumn 1978), 10-11.

Shank, Ellsworth B., "Navigation of the Lower Susquehanna," *Harford [County, Maryland] Historical Bulletin*, No 30 (Fall 1986), 84-99.

Shank, William H., *The Amazing Pennsylvania Canals* (York, Pa: American Canal and Transportation Center, 1960, 1973)
————————, "Canals of the Lower Susquehanna," *CC*, No. 59 (Summer 1982), 7.
————————, ed.), *The Columbia-Philadelphia Railroad and Its Successors* (York, Pa: American Canal and Transportation Center 1985).
————————, "Pennsylvania Canal Company (1857-1926)," *CC*, No. 73 (Winter 1986), 1-4, 15.
————————, *Three Hundred Years with the Pennsylvania Traveler* (York, Pa: American Canal and Transportation Center, 1976); revised with new title *Pennsylvania Transportation History*, 1990.

Sharf, John. Thomas and Thompson Westcott, *History of Philadelphia, 1609-1884*, 3 vols., (Philadelphia: L.H. Everts & Co, 1884).
————————, *History of Maryland from the Earlies Period to the Present Day*, in 3 volumes, (Hatboro, Pennsylvania: Tradition Press, 1967, reprint of 1879 edition), especially II, Chapter XXXI.
Sharp, Myron B, "Troubles on the Pennsylvania Canal," *WPHM*, LII (1969), 153-59.

Shaugnessy, Jim, *Delaware & Hudson: The History of an Important Railroad whose Antecedant was a Canal Network to Transport Coal* (Berkley, Calif: Howell-North Books, 1967).

Shoemaker, Henry W., "Last Pilot House on the Old Susquehanna Canal," *CC*, No 41 (Winter 1978), 11.

"A Side of Roebling that History has Forgotten," *CC*, No. 9 (Summer 1969), 3, 4.

Sioussat, St. George L., "Dr. William Smith, David Rittenhouse, and the Canal Plate September 7, 1777," *Proceedings of the American Philosophical Society*, LCV (1951), 223-231.

Smeltzer, Gerald, *Canals Along the Lower Susquehanna* (York, Pa: Historical Society of York County, 1961)

PENNSYLVANIA (Cont.).

Smith, Edwin F., "The Schuylkill Navigation," *Publications of the Historical Society of Schuylkill County,*" II (1910), 475-500; IV (1912),9-17, 353-380.

Snelling, Richard I., "Philadelphia and the Agitation in 1825 for the Pennsylvania Canal," *PMHB,* LXII (1938), 175-204.

Snow, Dr. F. Herbert, "Ohio River Navigation in Western Pennsylvania," *PDIA,* IX, # 3 (1941), 11-14; IX, # 4, 21-27.
——————————————, " Use and Navigation of Ohio River in Pennsylvania [Pymatuning Dam], " *PDIA,* IX # 7 (1941), 7, 23-28.

Snyder, Charles F., "The Pennsylvania Canal," *Proceedings of the Northumberland County Historical Society,* XXVI (1974), 44-59.

Soete, Henry G., "Notes on the D & H Canal," *CC,* No. 5 (Summer 1968), 5; No. 6 Fall 1968), 5; No 7 (Winter 1969), 2.
———————————, "John Wurts D & H 'Strong Man'," *CC,* No. 12 (Spring 1970), 8.

"Some account of the rise, progress and present state, of the Union Canal, of the state of Pennsylvania," *Journal of the Franklin Institute.* IV (1827), 274-277.

Speers, Sandra, "The Restoration of Roebling's Delaware and Hudson Aqueduct," *CCHT,* VI (1987), 113-146.

Stapleton, Darwin H., "Borrowed Technology on the Allegheny Portage Railroad," *CC,* No. 24 (Spring 1973), 6.
————————————, "Robert Brooke: Pennsylvania's First Canal Engineer?" *CC,* No. 35 (Summer 1976), 15.
————————————, "Solomon W. Roberts' Trip from Blairsville to Pittsburgh," *CC,* No 22 (Fall 1972), 3.
————————————, "Training of a Canal Engineer," *CC,* No. 28 (Spring 1974), 8.
————————————, "William Weston, Benjamin Henry Latrobe and the Philadelphia Plan for Improvements," in Steven Cutliffe (ed.), *Science and Technology in the Eighteenth Century* (Bethlehem, Pa., Lehigh Univ. Press, 1984).

"State's Industrial Greatness Traced in Part to Old Canals," *PDIA,* VII, #4 (1939), 3-13.

"State's Struggle for Supremacy Noted in its Early Canal History," *PDIA,* IV, # 1 (1937), 18-27.

PENNSYLVANIA (Cont.).

    Steers, Edward, "The Delaware and Hudson Canal Company's Gravity Railroad," *CCHT*, II (1983), 129-203.

    --------------- (contributor), " D & H Canal News 1863- 1871," *CC*, No 79 (Summer 1987), 4-8.

    ---------------, "The Pennsylvania Coal Company"s Gravity Railroad," *CCHT*, I (1982), 155-229+.

    Stephenson, Clarence D., "Civil War Excursion," *CC*, No. 8 (Spring 1969), 3.

    -----------------------, *The Pennsylvania Canal: Indiana and Westmorland Counties* (Marion Center, Pa: author, 1961, reprint, Indiana, Pa., Halldin, 1979), .

    Stowe, Mrs. H. E., "The Canal Boat," *CC*, No. 27 (Winter 1974), 5-6, [reproduced from *Godey's Lady's Book*, XXIII (4), October 1841, 167-169].

    Strickland, William, *Report on Canals, Railroads, Roads, and Other Subjects, Made to "The Pennsylvania Society for the Promotion of Internal Improvements*, (Philadelphia: Carey & Lea, 1826), also available on microfilm from the Pennsylvania Historical and Museum Commission [RG-17 Board of Canal Commissioners, General Records, Reel 8, # 3615].

The Sunbury Canal and Water Power Company, reproduction of the Brochure printed in 1842--with editorial notes by Charles Fisher Snyder, *Proceedings of the Northumberland County History Society*, XXIV (1963), 90-105.

    Swank, James M., *Progressive Pennsylvania; a Record of the Remarkable Industrial Development of the Keystone State, with Some Account of Its Early and Its Later Transportation Systems, Its Early Settlers, and Its Prominent Men* (Philadelphia: Lippencott, 1908).

    Swisher, Clare, "Exit: Erie Extension," *CC*, No. 55 (Summer 1981), 10, [reprinted from *The Erie Story*, September, 1967].

    Swetnam, George, *Pennsylvania Transportation*, (Gettysburg: Pennsylvania Historical Association, 1964, revised, 1968)

    Taber, Thomas T., "Bill Brobst and Bob Plankenhorn Reminisce on the Last Days of the Canal," *Now and Then*, XVI (1971), 702-705.

Tanner, H.S., *A Brief Description of the Canals and Rail Roads of Pennsylvania and New Jersey Comprehending Notices of all the Most Important Works of Internal Improvements in Those States* (Philadelphia: author, 1834).

------------, *Memoir of the Recent Surveys, Observations and Internal Improvements in the United States*, (Philadelphia: 1830).

------------ *The American Traveller or Guide through the United States*, (Philadelphia: author, 1840).

Tarr, Joel A., (ed.), "Philo E. Thompson's Diary of a Journey on the Main Line Canal," *PH*, XXXII (1965), 295-304; partially reprinted in *CC*, No. 79 (Summer 1987), 13-14.

"Tests on Old Portage Led to Use of Locomotives Instead of Horses," *PDIA*, VII, # 1 (1939), 16-31.

Theiss, Edwin Lewis, "Canallers," a chapter in George Korson (ed.), *Pennsylvania Songs and Legends* (Philadelphia: Univ. of Pennsylvania Press, 1949), 258-289.

------------------, "The Canals of Pennsylvania," *Proceedings of the Northumberland County Historical Society*, IV (1932), 67-90.

------------------, "Pirates along the Pennsylvania Canal," *Proceedings of the Northumberland County Historical Society*, XVI (1948), 102-117.

Thomas, George, "Proposed Canal From Baltimore to Conewago," *AC*, Bulletin No. 45 (May 1983).

Toogood, Anna Coxe, *Historic Resource Study--Allegheny Portage RR National Historic Site, Pennsylvania* (National Park Service, Denver, Colorado, May 1973).

Trautwine, John C. Jr., "The Philadelphia and Columbia Railroad of 1834," *Philadelphia History*, II (1925), 139-178

Trego, Charles B., *A Geography of Pennsylvania*, (Philadelphia: Edward C. Biddle, 1843)

[Trimble, Isaac Ridgeway], *Report of the Engineer Appointed by the Commissioners of the Mayor and city Council of Baltimore on the Subject of the Maryland Canal* [Susquehanna and Tidewater] (Baltimore: Lucas & Beaver, 1837).

"Tunnel Under Pittsburgh Joined Western Division with Monongahela River," from *HR*, XV (March 1835), 152; *CC*, No. 43 (Summer 1978), 1-2.

PENNSYLVANIA (Cont.)

"Union Canal Data Vast Historical Reservoir," *PDIA*, XVI, # 5 (1948), 11-13.

Unrau, Harlan D., *Historic Structure Report, Historic Data Section: The Delaware Aqueduct, Upper Delaware Scenic and Recreational River* (Denver, Colo: National Park Service, 1983).

Unrau, Harlan D. and Sandra Hauptman, "Roebling's Delaware Aqueduct During the 20th Century," *CCHT* III (1984), 135-172.

Van Fossen, W.H., "Sandy and Beaver Canal," *CC*, No. 47 (Summer 1979), 10--13 [from *Ohio State Archaeological and Historical Quarterly*], LV (April-June, 1946), 165-177].

Veech, James, *A History of the Monongahela Navigation Company* (Pittsburgh: 1873).

Vogel, Robert M., *Roebling's Delaware & Hudson Canal Aqueducts*, (Washington: Smithsonian, 1971, reprint 1984).

Vogel, Robert M and Sandra Hauptman, " Roebling's Delaware Aqueduct," *CC*, No. 61 (Winter 1983), 7-13.

Wakefield, Manville B., "The Delaware & Susquehanna Canal: A Canal that Never Got off the Drawing Boards," *AC, No. 12 (February 1975), 5.*
------------------------, *Coal Boats to Tidewater* (Grahamsville, N.Y: Wakefair Press, 1965, 1971)

Waltman, Charles, "The Influence of the Lehigh Canal on the Industrial and Urban Development of the Lehigh Valley," *CCHT* II (1983), 87-104.

Walton, Denver, "Guide to the Beaver Division Canal," *CC*, No. 34 (Spring 1976), 9-10.
---------------, "Youghiogheny Navigation," *CC*, No. 46 (Spring 1979), 4, 16.

Walton, Richard C., "Boyhood Days on Old Canal," *CC*, No. 68 (Autumn 1984), 12 [from *Harrisburg Patriot*, 1941].
------------------, "Bray of Old Gray Mule Canal Boat Whistle," *CC*, No. 58 (Spring 1982), 8 [from *Harrisburg Patriot*, 1940].
------------------, "Canal Camping Recalled," *CC*, No. 59 (Summer 1982), 6, [from *Harrisburg Patriot*, 1941].

PENNSYLVANIA (Cont.).

Walton, Richard C., "Canaller Recalls Flood," *CC,* No. 48 (Autumn 1979), 15 [from *Harrisburg Patriot,* 1940].

————————————, "Canal Rules Passed to Resolve Disputes," *CC,* No. 42 (Spring 1948), 6-7 [from *Harrisburg Patriot* in 1940s].

————————————, "Children on Canal Boats Liked to Pass Rockville," *CC,* No. 51 (Summer 1980), 12 [from *Harrisburg Patriot*].

————————————, "Dredgemen Feared PCC Closing," *CC,* No.73 (Winter 1986), 16 [from *Harrisburg Patriot,* 1941].

————————————, "Ever Try Sleeping on a Mule," *CC,* No. 49 (Winter 1980), 16 [from *Harrisburg Patriot,* 1941].

————————————, "To Huntingdon and Back in 1883," *CC,* No. 71 (Summer 1985), 13, 16 [from the *Harrisburg Patriot,* 1940s].

————————————, "Idea of Canals in Panna. Fostered by William Penn," *CC,* No. 43 (Summer 1978), 11-12 [from *Harrisburg Patriot,* 1941].

————————————, "Immigrants Used Penna Canals on Way Westward," *CC,* No. 65 (Winter 1984), 9 [from *Harrisburg Patriot,* 1941].

————————————, "Proponents of Rival Canal Routes Waged Long Fight," *CC,* No. 36 (Autumn 1976), 9 [from *Harrisburg Patriot,* 1940]..

————————————, "Quick Wit Wins Pot Pie," *CC,* No. 39 (Summer 1977), 11 [from *Harrisburg Patriot,* 1940

————————————, "632 Miles of Canals and RR's by 1834," *CC,* No. 69 (Winter 1985), 8 [from *Harrisburg Patriot,* 1941].

————————————, "Tunnel Considered for a Pa. Canal Route," *CC,* No. 64 (Autumn 1983), 4 [from the *Harrisburg Patriot,* 1941].

————————————, "Union Canal Put Up Stiff Battle to Escape Sheriff," *CC,* No. 55 (Summer 1981), 7 [from 1940s *Harrisburg Patriot*].

Ward, Isaac M., "Early Trip to Pittsburg," *AC,* No. 41 (May 1982), 6-7 [1836 Diary].

Watkins, J. Elfret, "History of the Pennsylvania Railroad Company, 1846-1896," three volumes (never published but bound copies of page proofs with engravings are in the Smithsonian Institution, Washinton, D.C. and in the Bureau of Railroad Economics Library, Washington, D.C., xerox copies at Center for Canal History and Technology, Easton, Pa.).

Watts, Irma A., "Pennsylvania Lotteries of Other Days," *PH,* II (1935), 36-43.

PENNSYLVANIA (Cont.).

    Wheeler, George, "Benjamin Aycrigg Pioneer in Canal Survey and Building," *PDIA*, VIII, # 3 (1940), 3-6.

    ----------------, "A Journey on the Union Canal and Schuylkill Navigation," *Historical Review of Berks County*, IV (1939), 101-103.

    ----------------, "The Union Canal and its Relation to Philadelphia," *City History Society of Philadelphia, Publications*, IV (1939), 75-91.

    Wheelersburg, Robert, "Archaeological Resources of 'Hinterland' Canal Communities in East Central Pennsylvania: The Bellfonte Project," *CCHT*, VIII (1989), 104-139.

    Whelan, Frank, "That Oily Herrenhutter: John Rice, The Jacksonian Era and the Collapse of the Northampton Bank, 1820-1861," *CCHT*, IX (1990), 55-81.

    Whippo, Charles T., "Engineer's Report on the Beaver Division," *CC*, No. 39 (Summer, 1977), 5.

    White, Josiah, "White Rebuts LC&N Critics," *CC*, No. 72 (Fall 1985), 6-9, 12.

    Whitenight, Hazel, "The Pennsylvania Canal," *PH*, X (1943), 297-299.

    Whyte, Larry E., " Phoenix Branch Canal," *CC*, No. 78 (Spring 1887), 7-9, 16.

    Williams, David G., "The Lehigh Canal System," *Proceedings of the Lehigh County Historical Society*, XXII (1958), 99-135.

    -------------------, "Transportation Thru the Lehigh Water Gap," *Proceedings of the Lehigh County Historical Society*, XXVI (1966), 257-265.

    Wilson, H. C., "The Schuylkill Canal," *Schuylkill County Historical Society, Publications*, IV (1914), 353-380.

    --------------, "Schuylkill Pioneer in Water Transportation," *PDIA*, XVIII, # 2 (1950), 8-15; XVIII, # 3, 11-16; XVIII, # 4, 18-23; XVIII, # 5, 13-18.

    Wilson, William Bender, *From the Hudson to the Ohio* (Philadelphia: Kensington Press, 1902).

    ----------------------, *History of the Pennsylvania Railroad*, vol 1, (Philadelphia: H. T. Coates and Company, 1899).

    ----------------------, *The Evolution, Decadence and Abandonment of the Allegheny Portage Railroad*, Annual Report of the Secretary of Internal Affairs, Part 4, Harrisburg, 1898-1899.

PENNSYLVANIA (Cont.)

Wilson, William Hasell, "Notes on the Columbia and Philadelphia Railroad, 1940," *Journal of the Franklin Institute*, XXIX (1840), 333-341.

----------------------, *Notes on the Internal Improvements of Pennsylvania* (Philadelphia: Railway World, 1879).

----------------------, *Reminiscences of a Railroad Engineer* (Philadelphia: Railway World Publishing Co., 1896).

"Wire Ropes on the Inclined Planes of the Allegheny Portage Railroad," *Journal of the Franklin Institute*, XXIX (1845), 212-213

"The Wire Suspension Aqueduct over the Allegheny River, at Pittsburgh," *Journal of the Franklin Institute*, XL (1845), 306-309.

Wistar, Isaac James, "Wistar Canal President," [from Autobiography *CC*, No. 73 (Winter 1986), 12-15.

Woodley, Thomas F., *Thaddeus Stevens*, (Harrisburg, Pa: Stackpole Sons, Publishers, 1937).

"Woodruff's Portage Journey," *CC*, No. 27 (Winter 1974), 1-2, 7 [from unpublished "Journal History of the Church of Jesus Christ of Latter Day Saints"].

Woods, Terry K., "Anthracite and Slackwater," *CCHT*, II (1983), 45-68.

----------------, "Canalling Coal to Erie," *CC*, No. 54 (Spring 1981), 8-9, reprint from *Erie Story Magazine*.

----------------, "French Creek Feeder," *CC*, No. 34 (Spring 1976), 6.

----------------, "Hanover and the Sandy - Beaver Canal," *CC*, No. 47 (Summer 1979), 1-3.

----------------, "The Horse Race," *CC*, No. 60 (Autumn 1982), 1, 7, [North Branch].

----------------, "The Junction Canal," *AC*, No. 50 (August 1984) 4-5.

----------------, "The North Branch Canal," *CC*, No. 70 (Spring 1985), 10-11.

----------------, "Pennsylvania North Branch Extension Canal," *AC*, No.25 (May 1978), 3.

----------------, "Sandy and Beaver Canal Short-Lived," *CC*, No. 16 (Spring 1971), 3.

----------------, "The Wyoming Division of the North Branch Canal," *CC*, No. 41 (Winter 1978), 4, 10.

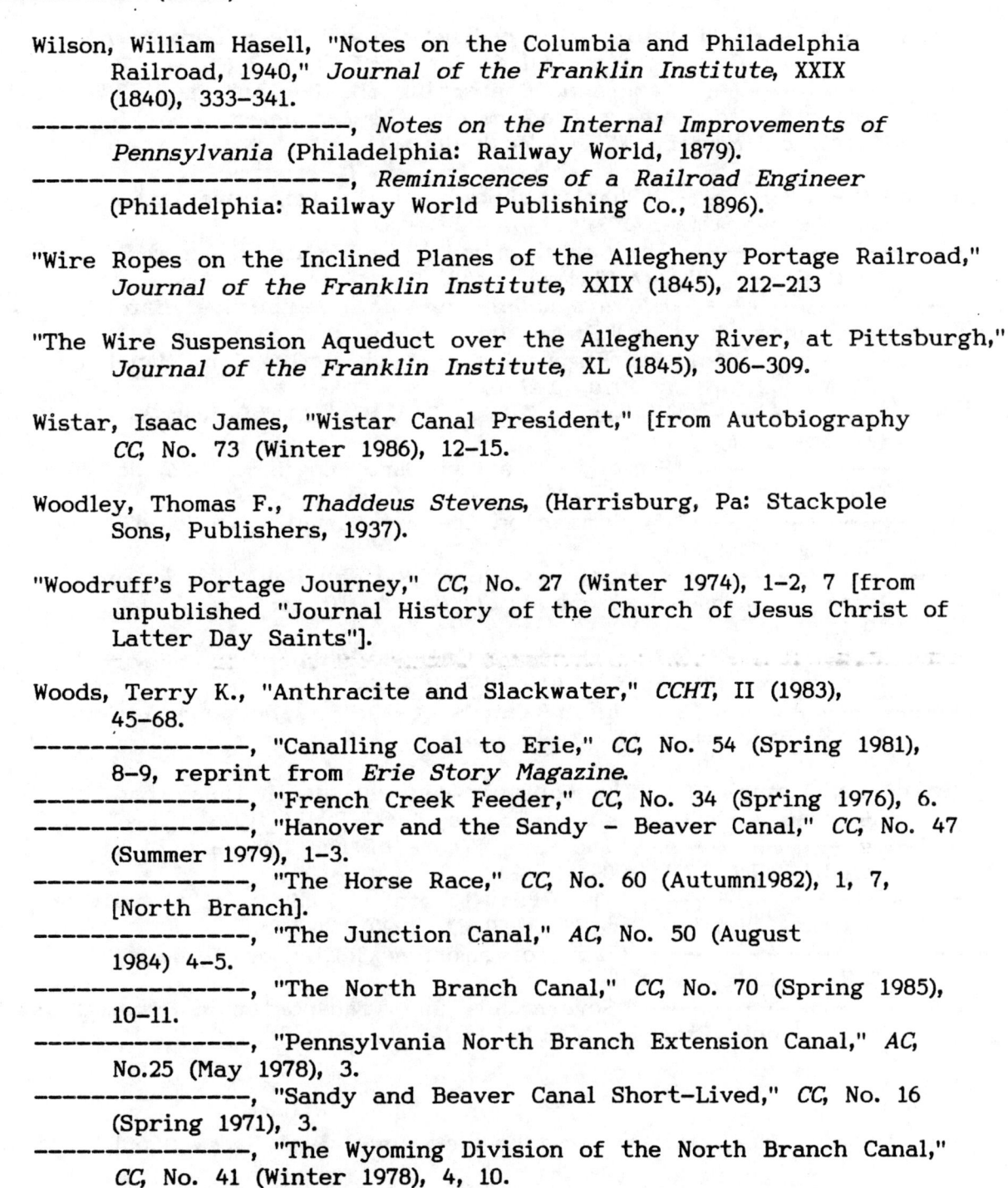

Worner, William F., "Riding on the Columbia and Philadelphia Railroad," *Papers Read before the Lancaster County Historical Society*, XXXV (1931), 89-95.

PENNSYLVANIA (CONT.)

Wright, Samuel C., "Paradise, O Paradise," *Papers Read before the Lancaster County Historical Society*, XVIII pt. 2 (1914), 42-51.

——————————— "Memoranda Concerning the Columbia and Philadelphia Railroad, Etc., *Papers Read before the Lancaster County Historical Society*, XXI (1917), 8-10.

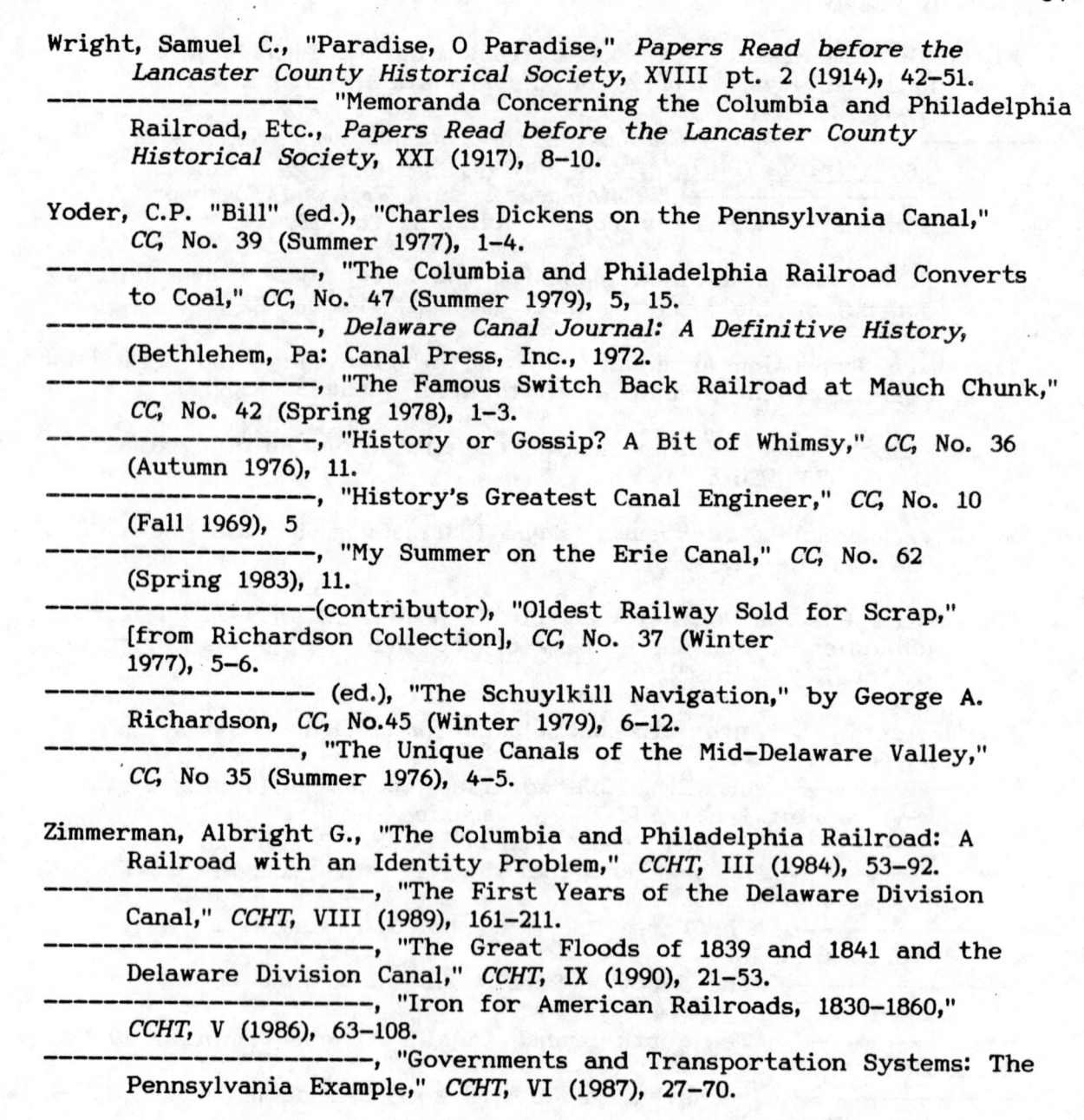

Yoder, C.P. "Bill" (ed.), "Charles Dickens on the Pennsylvania Canal," *CC*, No. 39 (Summer 1977), 1-4.

———————————, "The Columbia and Philadelphia Railroad Converts to Coal," *CC*, No. 47 (Summer 1979), 5, 15.

———————————, *Delaware Canal Journal: A Definitive History*, (Bethlehem, Pa: Canal Press, Inc., 1972.

———————————, "The Famous Switch Back Railroad at Mauch Chunk," *CC*, No. 42 (Spring 1978), 1-3.

———————————, "History or Gossip? A Bit of Whimsy," *CC*, No. 36 (Autumn 1976), 11.

———————————, "History's Greatest Canal Engineer," *CC*, No. 10 (Fall 1969), 5

———————————, "My Summer on the Erie Canal," *CC*, No. 62 (Spring 1983), 11.

———————————(contributor), "Oldest Railway Sold for Scrap," [from Richardson Collection], *CC*, No. 37 (Winter 1977), 5-6.

——————————— (ed.), "The Schuylkill Navigation," by George A. Richardson, *CC*, No.45 (Winter 1979), 6-12.

———————————, "The Unique Canals of the Mid-Delaware Valley," *CC*, No 35 (Summer 1976), 4-5.

Zimmerman, Albright G., "The Columbia and Philadelphia Railroad: A Railroad with an Identity Problem," *CCHT*, III (1984), 53-92.

———————————, "The First Years of the Delaware Division Canal," *CCHT*, VIII (1989), 161-211.

———————————, "The Great Floods of 1839 and 1841 and the Delaware Division Canal," *CCHT*, IX (1990), 21-53.

———————————, "Iron for American Railroads, 1830-1860," *CCHT*, V (1986), 63-108.

———————————, "Governments and Transportation Systems: The Pennsylvania Example," *CCHT*, VI (1987), 27-70.

FICTION

Fall, Thomas, *Canal Boat to Freedom* (New York: Dial Press, 1966).

PENNSYLVANIA (cont.)

JUVENILE

Leadership Lehigh Valley Class of 1990, *Lehigh Valley and Lehigh Canal: Cradle of the Industrial Revolution* (Joint Venture of Allentown-Lehigh County Bethlehem Area, Two Rivers Chambers of Commerce, 1990), coloring book.

UNPUBLISHED MANUSCRIPTS

MA Dissertations

Bastoni, Gerald Robert, "Canvass White, Esquire (1790-1834): Civil Engineer," MA, Lehigh University, 1983.

Corkan, Lloyd, "The Beaver and Erie Canal," MA, University o Pittsburgh, 1927.

Filippeli, Ronald L, "The Schuylkill Navigation Company and Its Role in the Development of the Anthracite Coal trade and Schuylkill County, 1815-1845," MA, Pennsylvania State Univ., 1966.

Kimes, Pearl Clair, "The Union Canal of Pennsylvania, a Thesis in History," MA, University of Pennsylvania, 1934.

Landsrath, Mary Hill, "A History of the Canal Movement in Pennsylvania," MA, Pennsylvania State Univ., 1917.

Kimes, Pearl C., "The Union Canal of Pennsylvania," MA, University of Pennsylvania, 1934.

Miller, Robert E., "The Sale of the Main Line of the Pennsylvania Public Improvements," MA, Pennsylvania State Univ., 1967.

Pratt, Elizabeth Miller, "The Building of the Pennsylvania Canal," MA, Columbia Univ., 1949.

Quinn, Sister Mary Faber, "The Influence of the North Branch Canal on the Pattern of Urban Land Use in Wilkes-Barre, Pennsylvania," MA, Catholic University of America, 1959.

PENNSYLVANIA (cont.)

PhD Dissertations

Brzyski, Anthony, "The Lehigh Canal and its Effect on the Economic Development of thr Region Through which it Passed, 1818-1873," (New York University, 1957).

Livengood, James W., "The History of the Commercial Rivalry between Philadelphia and Baltimore for the Trade of the Susquehanna Valley, 1780-1860," (Princeton, 1937).

Powell, H. Benjamin, "Coal, Philadelphia, and the Schuylkill," (Lehigh University, 1968)

Rubin, Julius, "Imitation by Canal or Innovation by Railroad: a Comparative Study of the Response to the Erie Canal in Boston, Philadelphia, and Baltimore," (Columbia, 1959)

Sadov, Abraham, "Transportation and the Appalachian Barrier: a Case Study in Economic Innovation," (Harvard, 1950)

Shegda, Michael, "History of the Lehigh Coal and Navigation Company to 1840," (Temple University, 1952).

Wallner, Peter A., "Politics and Public Works: Study of the Pennsylvania Canal System, 1825-1857," (Pennsylvania State University, 1973)

NEW JERSEY:

Bibliographical Aid

Hasse, Adelaide R., *Index of Economic Material in Documents of the States of the United States: New Jersey, 1789-1904* (Washington: Carnegie Institution, 1914, Kraus Reprint, 1965), "Canals,~ pps. 186-190

Delaware and Raritan Canal

Cawley, James and Margaret, *Along the Delaware and Raritan Canal*, (Rutherford, NJ: Fairleigh Dickinson University Press, 1970).

Cuyler, Lewis B., "Origins of the Delaware and Raritan Canal," *Princeton History*, 1983, No. 4, 1-16.

Davison, Betty B., *The Delaware and Raritan Canal: A User's Guide* (Princeton, N.J: The Delaware and Raritan Canal Coalition, 1976).

Delaware & Raritan Canal: Cadwalader Park Study, Trenton, N.J. (August, 1975).

Delaware and Raritan Canal State Park: Master Plan James C. Amon, preparer). (May, 1977); second edition (May 1989).
Delaware and Raritan Canal State Park: Design Guide (December, 1980).
Delaware and Raritan Canal State Park: Historic Structures Survey (June, 1982).

Godfrey, Capt. Frank H, "Notes on the Delaware and Raritan Canal," *CC,* No. 23 (Winter 1973), 3-4.

Howard, Henry, *The Yacht "Alice"; A Cruise from New York to Miami by Alice Sturtevant Howard* (Boston: Charles E. Lauriat Co, 1926).

Howell, J. Roscoe, "Ashbel Welch, Civil Engineer," *Proceedings of the New Jersey Historical Society*, Part I, LXXIX (1961), 251-263, Part II, LXXX (1962), 46-53; reprint, Lambertville Historical Society, 1973.

Knox, Nancy, "Princeton Basin," *Princeton History*, 1983, No. 4, 17-28.

McKelvey, William B., Jr., *The Delaware & Raritan Canal: A Pictorial History* (York, Pa: Canal Press, Inc., 1975).
---------------------- (ed.), "Operating Canal Locks by Steam," by Ashbel Green, original in *American Society of Civil Engineers Transactions,* IX, No. CXCIX, August 1880, *CC,* No. 69 (Winter 1985), 15-16.

NEW JERSEY–(cont.)

Delaware and Raritan Canal

Madeira, Crawford C., Jr., *The Delaware and Raritan Canal, A History* (East Orange, N.J: The Easterwood Press, 1941).

Menzies, Elizabeth G.C., *Millstone Valley* (New Brunswick: Rutgers Univ. Press, 1969).
——————————————, *Passage Between Rivers* (New Brunswick: Rutgers Univ Press, 1976)..

Neilson, James, "The Delaware and Raritan Canal: Some Early Recollections," *Proceedings of the New Jersey Historical Society*, LIII (1935), 131–132.

Reardon, Pat, "Architecture Along the D & R Canal," *New Jersey Outdoors*, XI(3) (May/June 1984), 16–18 [also followed by an illustrated D & R insert].

Shippen, Edward, "Reminiscences of Admiral Edward Shippen," *PMHB*, LXXVIII (1954). 201–230.

Sliney, Diane Jones, "Migrating South on the Delaware and Raritan," *AC*, No. 44 (February 1983), 8–9 [From *Time Off*, supplement to the *Princeton Packet*, November 3, 1981].

Smith, F. Hopkinson and J. B. Millet, "Snubbin' Thro' Jersey," *The Century Magazine*, XXXIV (1887), (new series XII) 483–496, 697–711, reprinted in pamphlet form by Canal Press, Inc., York, Pa., August, 1974 bound with Lawrence W. Pitt, "The Delaware and Raritan Canal: A Brief History."

Terhune, Laura P., *Episodes in the History of Griggstown: A Bicentennial Tribute Covering 300 Years* (New York: Albert H. Vela Co., 1976), Ch XI, "Mules and Barges," 74–85.

Thompson, Robert T., *Colonel James Neilson: A Business Man of the Early Machine Age in New Jersey, 1784–1862* (New Brunswick: Rutgers University Press, 1940).
——————————————, "Transportation Combines and Pressure Politics in New Jersey––1833–1836," *Proceedings of the New Jersey Historical Society*, LVII (1939), 1–15, 71–86.

Watkins, J. Elfreth, *The Camden & Amboy Railroad: Origin & Early History* (Washington, D.C: Press of Gedney and Roberts, 1891).

Welch, Ashbel, "Ship Canal Locks Calculated for Operation by Steam," *Transactions of the American Society of Civil Engineers*, IX (1880), 293–314.

NEW JERSEY (cont)

Morris Canal

 Chevalier, Michel, "Port Warren Plane in 1835," from *Vos de Communications Etats-Unis et des Traveox d'art qui en Dependant*, 1841-2, *CC*, Nos. 76-7 (Fall-Winter 1986-7, 7-14.

 Cranmer, Horace Jerome, "Internal Improvements in New Jersey: Planning the Morris Canal, (1822-24)" *Proceedings of the New Jersey Historical Society*, LXIX (1951), 324-341.

 Ferraro, William N., "Biography of a Morris Canal Village: Bowerstown, Washington Township, Warren County, New Jersey 1820-1940," *CCHT*, VIII (1989), 3-73.

 Goule, Frederick H., "The Decadence of an Old Canal," *CC*, No. 27 (Spring 1974), 5-6; No. 29 (Winter 1974), 5-6; No. 30 (Fall 1975), No. 31 (Winter 1975), [reproduced from *The Booklovers Magazine*, V (6), June 1905, 824-835.

 Hanson Kenneth R., "An Aerial Survey of the Remains of the Morris Canal," *Proceedings of the New Jersey Historical Society*, LXXXI (1963), 10-18.

 Heydinger, Earl J., "Financial Involvements of the Morris & Essex [sic.] Canal," *CC*, No. 49 (Winter 1980), 6-7.

 "Hydraulic Lift on the Morris Canal," reproduced with illustrations from *Scientific American*, XLVI, May 20, 1882, *CC*, Nos. 76-77 (Fall-Winter 1986-87), 6.

 Kalata, Barbara N., *A Hundred Years, A Hundred Miles, New Jersey's Morris Canal*, (Morristown, NJ: Morris County Historical Society, 1983)

 Lane, Wheaton J., "The Morris Canal," *Proceedings of the New Jersey Historical Society*, LV (1937), 214-231, 251-263.

 Lawson, Carl W. (contributor), "Inclined Planes of the Morris Canal," [from *The American Engineer and Railroad Journal*, Dec. 1894], *AC*, No. 37 (May 1981), 6-7.

 "Lawsuit Eyes Life on Canals," *CC*, Nos. 76-7 (Fall-Winter 1986-7), 15-20, 24.

 Lee, James, "The Morris Canal-A Brief History," *CC*, Nos. 76-77 (Fall-Winter 1986-87), 4-5, 24.
 ----------, *The Morris Canal: A Photographic History* (York, Pa: Canal Press, Inc, 1973), 2d ed, 1974, 3d revision, (Easton, Pennsylvania: Delaware Press, 1979), 4th revision, 1983.
 ----------, *Tales the Boatmen Told* (Exton, Pa: Canal Press, Inc., 1977).

NEW JERSEY (Cont.)

Morris Canal

Lee, James and Joe Hannan, "Morris Canal and Its People," *New Jersey Outdoors*, XV(3) (May/June 1988), 8-11.

Map and Illustrations of the Morris Canal Water Parkway, (Montclair, NJ: Morris Canal Parkway Association, 1914, reprinted, 1975).

Morrell, Brian H. (preparer), *Historic Preservation Survey of the Morris Canal in Warren County, New Jersey*, [prepared, September 1983], (Warren County Morris Canal Committee, 1987).

"Morris Canal Inventory 1892," *CC*, Nos. 76-77 (Fall-Winter 1986-87), 21-4.

"Morris Canal Plane Blueprints," reproduced from Michel Chevalier, *CC*, No 53 (Winter 1981), 8-9; also in Nos 76-7 (Fall-Winter 1986-7), 15-6.

Moss, Bill, "New Jersey's Unique Morris Canal," *CC*, No. 53 (Winter 1981), 1-2.

Munroe, John A., *Louis McLane: Federalist and Jacksonian* (New Brunswick: Rutgers Univ. Press, 1973) [Morris Canal].

Stickel, Fred G., "Through the Morris Canal," *New Jersey History*, LXXXIX (1971), 93-114).

Vermeule, Cornelius C. Jr., "The Morris Canal," *The Newcomen Society Transactions* (English), XV (1934-1935), 195-202.
—————————————, *State of New Jersey Morris Canal and Banking Company Final Report of Consulting and Directing Engineer* [historical sketch, pp. 49-80] (Trenton, N. J: Morris Canal and Banking Company, 1929).

Yoder, C.P., "Early Inclined Planes on the Morris Canal," *CC*, No. 50 (Spring 1980), 5, 15.
—————————, "The 'High Climbing' Morris Canal," *CC*, No. 15 (Winter 1971), 4.

Other Works:

Boyer, Charles S, *Waterways of New Jersey: History of Riparian Ownership and Control Over the Navigable Waters of New Jersey* (Camden, N.J: Sinnickson Crew & Sons Company, 1915).

Cadman, John W., *The Corporation in New Jersey: Business and Politics, 1761-1875* (Cambridge, Mass: Harvard Univ. Press, 1949).

NEW JERSEY (Cont.)

Other Works

    Cawley, James and Margaret, *Exploring the Little Rivers of New Jersey*, (Princeton: Princeton Univ. Press, 1942, 3d ed., 1971).

    Cranmer, H. Jerome, *The New Jersey Canals: State Policy and Private Enterprise 1820-1832* (New York: Arno, 1978), reprint of PhD Dissertation.

    "Early Memorial [1846] About a 'Ship Canal' in New Jersey," *Proceedings of the New Jersey Historical Society*, ns XIII (1928), 429-430.

    Haupt, Lewis M., "The Proposed Ship Canal Between New York and Philadelphia Connection the Delaware and Raritan Rivers," *Journal of the Franklin Institute*, CXXXIII (1892), 172-175 [with map].

    Lane, Wheaton J., *From Indian Trails to Iron Horse: Travel and Transportation in New Jersey, 1620-1860* (Princeton, N.J: Princeton Univ. Press, 1939).

    Johnson, James P., *New Jersey: History of Ingenuity and Industry* (n.p: Windsor Publications, Inc., 1987)

    Lenik, Edward J., "The Tuxedo-Ringwood Canal," *Proceedings of the New Jersey Historical Society*," LXXXIV (1966), 271-273.

    "Proposed New Jersey Canal of 1826," *Proceedings of the New Jersey Historical Society*, ns XIV (1929), 203-213.

    Rankin, Edward S., "Proposed Early Ship Canal Across Newark Meadows," *Proceedings of the New Jersey Historical Society*, ns XV (1930), 351-360.

    Vermeule, Cornelius C., "Early Transportation in and about New Jersey," *New Jersey Historical Society Proceedings*, IX (1924), 106-124.

    Viet, Richard F., *The Old Canals of New Jersey: A Historical Geography* (Little Falls, N.J: New Jersey Geographical Press, 1963).

NEW JERSEY (cont.)

Juvenile:

Dorian, Edith M., *High Water Cargo* [Along the Delaware and Raritan Canal, 1854], (New Brunswick: Rutgers University Press, 1950-1965).

Homer, Larona, "Life on the Canal," in Larona Homer, *The Shore Ghosts and Other Stories of New Jersey* (Wallingford, Pennsylvania: Middle Atlantic Press, 1981), 117-133.

Thompson, Mary W., *A Summer's Adventure on the Morris Canal* (Roxbury Township Historical Society, 1974).

Unpublished Mss

Cadman, John W., "The Policy of New Jersey with Respect to Corporations," 2 vols, PhD, Princeton Univ., 1947.

Cranmer, H. Jerome, "New Jersey Canals: a Study of the Role of Government in Economic Development," PhD, Columbia, 1955.

Froment, F. L., "The New Jersey Canals," Thesis, Princeton, 1931.

Pitt, Lawrence W. and Catherine B Pitt (researchers, compilers and collaters), "An Index of Historical Materials on the Planning to Construction of the Delaware and Raritan Canal from the Personal Papers aof Canvass White, Chief Engineer", from Department of Manuscripts and University Archieves, Cornell University, xerox copy in Museum Support Center, Hugh Moore Historical Park and Canal Museum, Easton, Penna.

Reilly, George L. A., "The Camden and Amboy Railroad in New Jersey Politics, 1830-1873," PhD, Columbia, 1950.

NEW YORK

Bibliographical Aids

Hasse, Adelaide R., *Index of Economic Materials in Documents of the States of the United States: New York, 1789-1904* (Washington: Carnegie Institution, 1907, Kraus Reprint, 1965)), "Canals," pps. 86-156.

Wyld, Lionel (comp.), *The Erie Canal: a Bibliography* (American Canal Society, 1978).

Other works

Albion, Robert Greenhaigh, *Rise of New York Port, 1815-1860* (New York: Scribner, 1939, reprint, Hamden Conn: Archon Books, 1961), [Chapter V, "Hinterland. . ."].

Anderson, Patricia, *The Course of Empire: The Erie Canal and the New York Landscape* (Rochester, NY: Memorial Art Gallery of the University of Rochester, 1984) [reproduction of canal paintings and bibliographies of literary views].

Andrist, Ralph K., *The Erie Canal* (New York: American Heritage Library, 1964).

Andrist, Ralph K. and Donald Plowden, "The Erie Canal Passed This Way," *American Heritage*, XIX (1968), 22-31, 77-80.

Archdeacon, Thomas F., "The Erie Canal Ring, Samuel J. Tilden and the Democratic Party," *New York History*, LIX (1978), 408-429.

Beame, Edmund M., "Rochester's Flour Milling Industry in the Pre-Canal Days," *Business History Review*, XXXI (1957), 209-225.

Beyer, Barry K., *The Chenango Canal* (New York: American Heritage Library, 1954, 1964; Norwich, NY: Chenango County Historical Society, 1968).

Brong, Karl S., "Genesis of the Erie Canal," *National Republic*, XXII (May 1934), 21-22, 30.

Brown, Mabel A., "Broadcasting By Canon," *National Republic*, XXI (April 1934), 20,32.

Brunger, Eric and Lionel Wyld, *New York's First Thruway* (Buffalo and Erie Historical Society, 1964).

"Buffalo and Black Rock Harbor Papers and Related Documents of Early Date," *Buffalo Historical Society Publications*, XIV (1910), 307-388.

NEW YORK–(cont.)

Canal Enlargement in New York State: Papers of the Barge Canal Campaign and Related Topics (Buffalo Historical Society Proceedings, XIII, 1909); "Canal Enlargement, 1-194; other canal items, 195-233, 303-308, 331-394.

Canal Society of New York, *Canals at Cohoes*, compiled by Thomas X. Grasso for Field Trip, October 4, 1980.
——————————————, *Canals of the Central Mohawk Velley*, compiled by Thomas X. Grasso for Field Trip, June 13, 1987.
——————————————, *Champlain Canal: Watervliet to Whitehall*, compiled by Thomas X. Grasso for Field Trip, October 5, 1985.
——————————————, *Erie Canal: Buffalo to Lockport*, compiled by Henry H. Baxter for Field Trip, October 3, 1987.
——————————————, *Genesee Valley Canal: Rochester to Dansville*, compiled by J. Hayward Madden for Field Trip, April 30, 1988.

"Canals in the State of New York, *Journal of the Franklin Institute*, XVIII (1834), 66-70.

Chalmers, Harvey, II, *The Birth of the Erie Canal* (New York: Bookman Associates, 1960.
————————————, *How the Irish Built the Erie Canal* (New York: Bookman Associates, 1964).

Cleland, C.E and L.M. Stone, "Archaeology as a Method for Investigating the History of the Erie Canal System," *Historical Archaeology*, I (1967), 63-70.

Chenango Canal, 1833-1878 (n.p: prepared by the Chenango County Planning Department, n.d.)

Clinton, George W., "Journal of Tour from Albany to Lake Erie by the Erie Canal in 1826," *Buffalo Historical Society Publications* (1910), XIV, 274-305.

Colden, Cadwallader D., *MEMOIR at the Celebration of the Completion of the New York Canals* (New York: 1825, reprinted, Ann Arbor, Michigan: University Microfilms, 1967).

Condon, George E., *Stars in the Water: The Story of the Erie Canal* (Garden City, N.Y: Doubleday & Co. Inc., 1974).

Cooke, Patricia, "The Erie Canal: American History Through Folklore," *New York Folklore*, V (1979), 155-167.

NEW YORK—(cont.)

    Cross, Whitney R., *The Burned-over District: The Social and Intellectual History of Enthusiastic Religion in Western New York, 1800-1850* (Ithaca: Cornell Univ. Press, 1950: reprint Harper Torchbook, 1965); Chapter IV, "Canal Days".

    Davidson, C., "Cruising by Packet Boat [Erie and Champlain Canal," *Americana*, IX (July/August 1981), 52-56.

    Dunn, Jamwa T. (ed.), "A Trip on the Northern Canal," *New York Folklore Quarterly*, VI (1950), 234-239.

    Ellis, David M., "Albany and Troy—Commercial Rivals," *New York History*, XXIV (1948), 268-300.
    ————————, "Rivalry Between the New York Central and Erie Canal," *New York History*, XXIX (1948), 268-300.

    The Erie Canal Centenary," *Buffalo Historical Society Proceedings*, XXII (1918), 269-295.

    Erie Canal Museum: Photos from the Collection (Syracuse, N.Y: Erie Canal Museum, 1989).

    Erie Canal Museum, *Homefront: The Erie Canal in the Civil War* (Syracuse, N.Y: Erie Canal Museum, 1987).

    "Erie Canal Passed This Way," *American Heritage*, XIX (October 1968), 22-31+.

    "Excavating Machines in the New York Barge Canal," *Engineering News*, LVII (June 6, 1907), 607, additional illustrations, pps. 608-609.

    "Failure of Masonry Arch Carrying the Erie Canal Over Onondaga Creek, Syracuse, N.Y.," *Engineering News*, LVIII (Aug 8, 1907), 151-152 [illustrations].

    Fairlie, John A., "Canal Enlargement in New York State," *Quarterly Journal of Economics*, XVII (1904), 286-292.
    ————————, "New York Canals," *Quarterly Journal of Economics*, XIV (1900), 212-239.
    ————————, "The New York Canals," *Annals of the American Academy of Political amd Social Science*, XXXI (1908), 117-125.

    Figliomeni, Michelle P., "The Canal That Never Was: The Orange and Sussex Canal," *Orange County Historical Society Publication No. 7, 1977-1978*, 17-32.

NEW YORK (cont.)

Finch, Roy G., *The Story of the New York State Canals, Historical and Commericial Information*, (Albany: J.B. Lyon Co., 1925.
———————, "John Bloomfield Jervis, Civil Engineer," *Transactions of Newcomen Society* [English]. XI (1930-1931), 109-120.

FitzSimons, Neal (ed.), *The Reminiscences of John B. Jervis, Engineer of the Old Croton* (Syracuse, NY: Syracuse Univ. Press, 1971).

Garrity, Richard G, *Canal Boatmen: My Life on Upstate Waterways* (Syracuse: Syracuse University Press, 1977).
———————, *Recollections of the Erie Canal* (Tonawanda, N.Y: Historical Society of the Tonowandas, 1966.

Godfrey, Capt. Frank H., *The Godfrey Letters* ed. Arnold H. Barben, (Syracuse: Canal Society of New York State, 1973).

Hanyan, Craig R., "China and the Erie Canal," *Business History Review*, XXXV (1961), 558-566.
———————, "China and the Erie Canal," iagara Frontier, XII (Autumn 1965), 71-77.

Haydon, Roger (ed.), *Upstate Travels: British Views of Nineteenth-Century New York* (Syracuse: Syracuse Univ. Press, 1982) section of selections on "Erie Canal".

Hawthorne, Nathaniel, "The Canal Boat," in "Sketches from Memory," *Mosses From an Old Manse*, (Century Edition of the Works of Nathaniel Hawthorn, Columbus, Ohio: Ohio State Univ. Press, 1974), X, 429-438.

Herndon, G. Melvin, "A Grandiose Scheme to Navigate and Harness Niagara Falls, *New-York Historical Society Quarterly*, LVIII (1974), 7-17.

Higbee, Elizabeth M., "They Remember the Erie Canal," *New York Folklore Quarterly*, XI (1955), 91-96.

Hill, Henry Wayland, "An Historical Review of Waterways and Canal Construction in New York State," *Buffalo Historical Society Publications*, XII (1908).
———————, "Historical Sketch of Niagara Ship Canal Project," *Buffalo Historical Society Publications*, XXII (1918), 208-266.

Holton, Gladys Reid, *The Genesee Valley Canal* (Brockport, N.Y: Stylus Graphics, 1970).

NEW YORK–(cont.)

Homefront: The Erie Canal in the Civil War (Syracuse, N.Y: Erie Canal Museum, 1987).

Hopkins, Vivian C., "The Governor and the Western Recluse: De Witt Clinton and Francis Adrian Van Der Kampe," *Proceedings of the American Philosophical Society,* CV (1961), 315–333.
——————————, "John Jacob Astor and DeWitt Clinton; Correspondence fron Jan. 25, 1808 to Dec. 23, 1827," *New York Public Library Bulletin,* LXVII (Dec. 1964), 654–673.

Hosack, David, *Memoir of DeWitt Clinton* (1829, reprinted, Ann Arbor: Michigan, University Microfilms, 1967).

Hotchkill, William O., *Early Days of the Erie Canal,* The Newcomen Society, American, 1940.

Howell, D.J., "Methods and Results of Surveys and Borings for Oswego-Mohawk Ship Canal Route," *Engineering News,* XLIII (June 21, 1900), 402–405; (June 28, 1900), 418–422.

Huber, William L. "History of the Sinnecock Canal," "How the Sinnecock Canal Works," No. 59 (Nov. 1986), *AC,* # 4, 5.

Hudowalski, E.C., "New York State Barge Canal System," *Proceedings of the American Society of Civil Engineers,* 1959, paper 2176.

Hulbert, Archer B., *The Great American Canals,* Vol. II, *The Erie Canal,* Vol 14, Historic Highways of America (Cleveland, Ohio: Arthur H. Clark Co., 1904).

Kent, Norman, "The Erie Canal, A Record in Pastels," *American Artist,* XXX (June, 1966), 42–47, 78–80.

Kimball, Francis P., *New York--the Canal State* (Albany: Argus Press, 1937)

Langbein, Walter. B., *Hydrology and Environmental Aspects of the Erie Canal (1817–1899),* Geological Survey Water-Supply Paper 2038 (Washington: Government Printing Office, 1976).
——————————, "Our Grand Erie Canal: 'a splendid project, a little short of madness,'" *Civil Engineering,* XLVII (1977), 75–81.

Lowe, Jack W. "Modeling an Erie Canal Packet," *Nautical Research Journal* XXI(3) (Sept 1975), 107–113.

McCombs, Hazel A., "Erie Canawl Lore," *New York Folklore Quarterly,* III (1947), 204–212.

NEW YORK (cont.)

McKee, Harley J., "Canvass White and Natural Cement, 1818-1825," *Society of Architectural Historians Journal*, XX (Dec 1961), 194-197.

McKelvey, Blake, "The Erie Canal, Mother of Cities," *New York Historical Quarterly*, XXXV (Jan. 1951), 55-71.
----------------, "Rochester and the Erie Canal," *Rochester History*, XI (July 1949), 1-24.

Mabie, Roger W., "The Hudson River Port of Rondout," *Sea History*, XXXVII (1985), 12-15.

Matthews, Lois Kimball, "The Erie Canal and the Settlement of the West," *Buffalo Historical Society Publications*, XIV (1910), 187-203.

Merrill, Arch, *The Towpath* (Rochester, N.Y: Gannett Company, Inc., 1945).

Miller, Nathan, *The Enterprise of a Free People: Aspects of Economic Development in New York State During the Canal Period, 1792-1828* (Ithaca, N.Y: Cornell University Press, 1962).
----------------, "Private Enterprise in Inland Navigation: The Mohawk River Prior to the Erie Canal," *New York History*, XXXI (1950), 398-413.

Mills, James Cooke, "Construction Work on the Erie Barge Canal," *Cassier's Magazine*, XXXIII (1907-1908), 636-649.

Moeller, Jan., "Chugging Along the Erie Canal," *Cruising World*, VII (Nov 1981), 82-84.

Moran, Eugene F., "The Erie Canal as I have Known It," *Bottoming Out* III, No.2 (1959), 1-18.

"The Mott Haven Canal," *AC*, No. 51 (November, 1984), 5, reprinted from *"Yankees" Magazine*, June 28, 1984.

"The New York Barge Canal and the Federal Deep Waterway: A Comparison," *Engineering News*, XLIX (Feb 26, 1903), 194-196.

"New York's Canal System," includes
William B. Shaw, "A Century's Growth," 53-56.
Henry W. Hill, "The Improved Barge Canals," 55-60,
Charles E. Ogden, "A Modern Canal Voyage from Lake Erie to the Hudson," 61-64, *American Review of Reviews*," XLVII (1923), 53-64.

"The Proposed Niagara Ship Canal," *Engineering News and American Railway Journal*, XXIII (Jan-June 1890), 434-435.

NEW YORK- (cont.)

North, Edward P., "Erie Canal and Transportation," *North American Review*, CLXX (1900), 121-133.

"Notes on the Service of Israel T. Hatch in Behalf of New York's Canals," *Buffalo Historical Society Proceedings*, XIV (1910), 389-396.

O'Brien, Charles F., "The Champlain Waterway, 1783-1897", *New England Quarterly,* LXI (June 1988), 163-182.

O'Donnell, Thomas C., *Snubbing Posts: An Informal History of the Black River Canal* (Boonville, N.Y: Black River Books, 1949), 2d edition (Old Forge and Lakemont, N.Y: North Country Books, 1972).

O'Donnell, Thomas F., "'I'm Afloat!' on the Raging Erie," *New York Folklore Quarterly*, XIII (1957), 177-180.

Phillips, Albert, *Along the Chenango Canal* (Norwich, N.Y: author, 1964).

Pritchard, Georgiana, "On the Erie Canal," *New York Folklore Quarterly*, X (1954), 45-47.

Pyle, Howard [depicted with pen and pencil by], "Through Inland Waters," *Harper's Magazine* XCII (1895), 828-839; XCIII (1896), 63-75.

Rauber, Wilfred J., *D[elaware]. & M[t. Morris]. and D[elaware]. L[ackawanna]. & W[estren]. Putting Danville on the Railroad Map. . . .with a Glance at the Genesee Valley Canal* (Dansville, N.Y: author, 1980).

Report of the Governor and the Legislature on the Overall Plan for Coordinated Lond-Range Development of Tourism and Economic Potential of New York's Canals, Prepared by The Barge Canal Planning and Development Board, February 1989.

Rezneck, Samuel, "Joseph Henry Learns Geology on the The Erie Canal in 1826," *New York History,* L (1969), 29-42.
----------------, "Travelling School of Science on the Erie Canal to 1825," *New York History,* XL (1959), 255-269.

Rideing, William H., "Water Ways of New York," *Harpers*, XLVIII (1873), 1-17.

Rinker, Harry L., *The Old Ragin Erie. . . There have been several changes* [A Postcard History], (Berkley Heights, N.J: Canal Captain's Press, 1984).

NEW YORK (Cont.)

Schell, Ernest H., "An Early Trip on the Erie Canal," *AC*, No. 28 (February 1979), 3-4.

Scherer, John L., *Greek Temples on the Towpath, A History and Guide to the Vischer Ferry Historic District* (Clifton Park, N.Y: Town of Clifton Park, 1977, 1986).

Sciaky, Leon, "The Roundout and Its Canal," *New York History*, XXII (July 1941). 272-289.

Seaton, Charles, "Gowanus Canal Future is 'Murky'!!" *AC* No. 57 (May 1986), 6-7.

Seelye, John, "'Rational Exultation': The Erie Canal Celebration," *Proceedings of the American Antiquarian Society*, XCIV (1984), 241-267.

Severance, Frank H. (ed.), "The Holland Land Co. and Canal Construction in Western New York" [correspondence], *Buffalo Historical Society Publications*, XIV (1910), 1-185.

Shaw, Ronald C., *Erie Water West: A History of the Erie Canal, 1792-1854* (Lexington, Ky: University of Kentucky Press, 1966, 1990).
—————————, "Michigan Influence Upon the Formative Years of the Erie Canal," *Michigan History*, XXXVII (1953), 1-18.

Shelton, Ronald L., *The New York State Barge Canal System* (Ithaca, N.Y: Cornell Univ. Resource Center, 1958).

Spier, Peter, *The Erie Canal* (New York: Doubleday & Co., Inc., 1970).

Smith, Walter B., "Wage Rates on the Erie Canal," *Journal of Economic History*, XXXIII (1963), 298-311.

Soule, C. F., *The Chenango Canal* (Canal Society of New York, 1970).

Staff of The Canal Museum, *A Canalboat Primer on the Canals of New York State* (Syracuse, N.Y: Canal Museum, 1981)

Stanley, Edward, "Waterway West," *Holiday* XV (June 1954), 98-101, 136-140, 143, 146.

State of New York, *The Erie Canal Centennial Celebration, 1926* (Albany: J.B. Lyon Co., Printers, 1928).

NEW YORK- (Cont.)

Stevens, John, *Documents Tending to Prove the Superior Advantages of Rail-Ways and Steam Carriages over Canal Navigation*, (New York: T & J Swords, 1812), Sanford & Swords, 1852, reprinted as extra No. 54, of *The Magazine of History with Notes and. Queries*, (Tarrytown: NY, 1917), reprinted Baker Library, Harvard Univ., 1935.

Stone, William Leete, "From New York to Niagara: Journal of a Tour, in Part by Canal in 1829," *Buffalo Historical Society Publications*, XIV (1910), 205-272.

Tarkov, John, "Engineering the Erie Canal, *American Heritage of Invention & Technology*, II (Summer 1986), 50-57.

Thomas, Phelan, *The Hudson Mohawk Gateway, An Illustrated History* (Northridge, Callifornia: Windsor Publications, 1985).

Thompson, Harold W., *Body, Boots & Britches* (Philadelphia et al: J.B. Lippencott Co., 1939); Chapter X, "Canawlers," 220-254.

Walsh, Edward S., *The Canal System of New York State* re-issued and revised, May 1, 1923 (Albany: J.B. Lyon co., 1923).

Veeder, David H., *The Original Erie Canal at Fort Hunter* (New York: Fort Hunter Canal Society, 1968).

Vogel, Robert M., (ed.), *A Report of the Mohawk-Hudson Area Survey conducted by the Historic American Engineering Record* (Washington, D.C: Smithsonian Institution Press, 1975)

Waggoner, Madeline Sadler, *The Long Haul West: The Great Canal Era, 1817-1850* (New York: G. P. Putnam's Sons, 1958).

Walker, Barbara K and Warren S., (eds.), *The Erie Canal: Gateway to Empire*, selected source materials for college research papers, (Boston: D.C. Heath, 1963).

Walsh, Edward S., *The Canal System of New York State*, re-issued and revised, May 1, 1923 (Albany: J.B. Lyon Co., 1923).

Watson, Elkanah, *History of the Rise, Progress, and Existing Conditions of the Western Canals in the State of New York from September, 1788, to the Completion of the Middle Section of the Grand Canal in 1819.* (Albany, D. Steele, 1820); microfiche edition, 1970; microfilm editions 1949, 1978 and 1980.

Watt, D.A., "The Siphon Lock on the New York Barge Canal at Oswego, N.Y.," *Engineering News*, LXIV (Nov 17, 1916), 530-533.

NEW YORK (CONT.)

Whitford, Noble E., "The Barge Canal Crossing of Oak Orchard Creek, Medina, N.Y.," *Engineering News*, LXX (July 13, 1913), 192-196; see also Emile Low, "Construction of the Barge Canal Crossing of Oak Orchard Creek," LXXIII (march 4, 1915), 430-432.

——————————, *History of the Canal System of the State of New York*, 2 vols. (Albany: supplement to the *Annual Report of the State Engineer and Surveyor for the State of New York for 1903-1905*, 1906)

——————————, *History of the New York Barge Canal of the State of New York* (Albany: supplement to the *Annual Report of the State Engineedr and Surveyor for 1921* State of New York, 1922).

Wiley, Day Allen, "The Enlargement of the Erie Canal," *Scientific American*, XCV (July 21, 1906), 45-46.

Williams, Emily, *Canal Country: Utica to Binghamton*, photography by Helen Cardemone (Utica, N.Y: E. Williams, c1982).

Williams, Mentor L., "Horace Greeley Tours the Great Lakes," *Inland Seas*, III (1947), 137-142.

Wilner, M.M., "The Erie Canal—Its Past and Future," *American Monthly Review of Reviews*, XXVIII (1903), 59-67.

Winslow, David, "Canal Diary," *New York Folklore Quarterly*, XVII (1961), 56-59.

Wyld, Lionel D., *Boaters and Broomsticks, Tales &Historical Lore of the Erie Canal* (Utica, N.Y: North Country Books, 1986).

—————————— (ed.), "A Farce on Erie Water," *New York Folklore Quarterly*, XVII (1961), 59-62.

——————————, *40'x28'x4'. The Erie Canal- 150 Years*, (Rome, NY: Oneida County Erie Canal Commemoration Commission, 1967)

——————————, *Low Bridge! Folklore and the Erie Canal* (Syracuse, N.Y: Syracuse University Press, 1962).

——————————, MrsErie Canal," *New York Folklore Quarterly*, XIV (1958), 265-268.

——————————, "Notes for a Yorker Dictionary of Canalese," *New York Folklore Quarterly*, XV (1959), 264-273.

Yates, Ray Francis, "How the New York Barge Canal Will be Operated," I, *Engineering News*, LXXV (Jan 20, 1916), 98-104; II, (Jan 27, 1916), 158-162.

——————————, "New York State Barge Canal," *Scientific American*, CXI (1914), 492-493.

NEW YORK (cont.),

FICTION

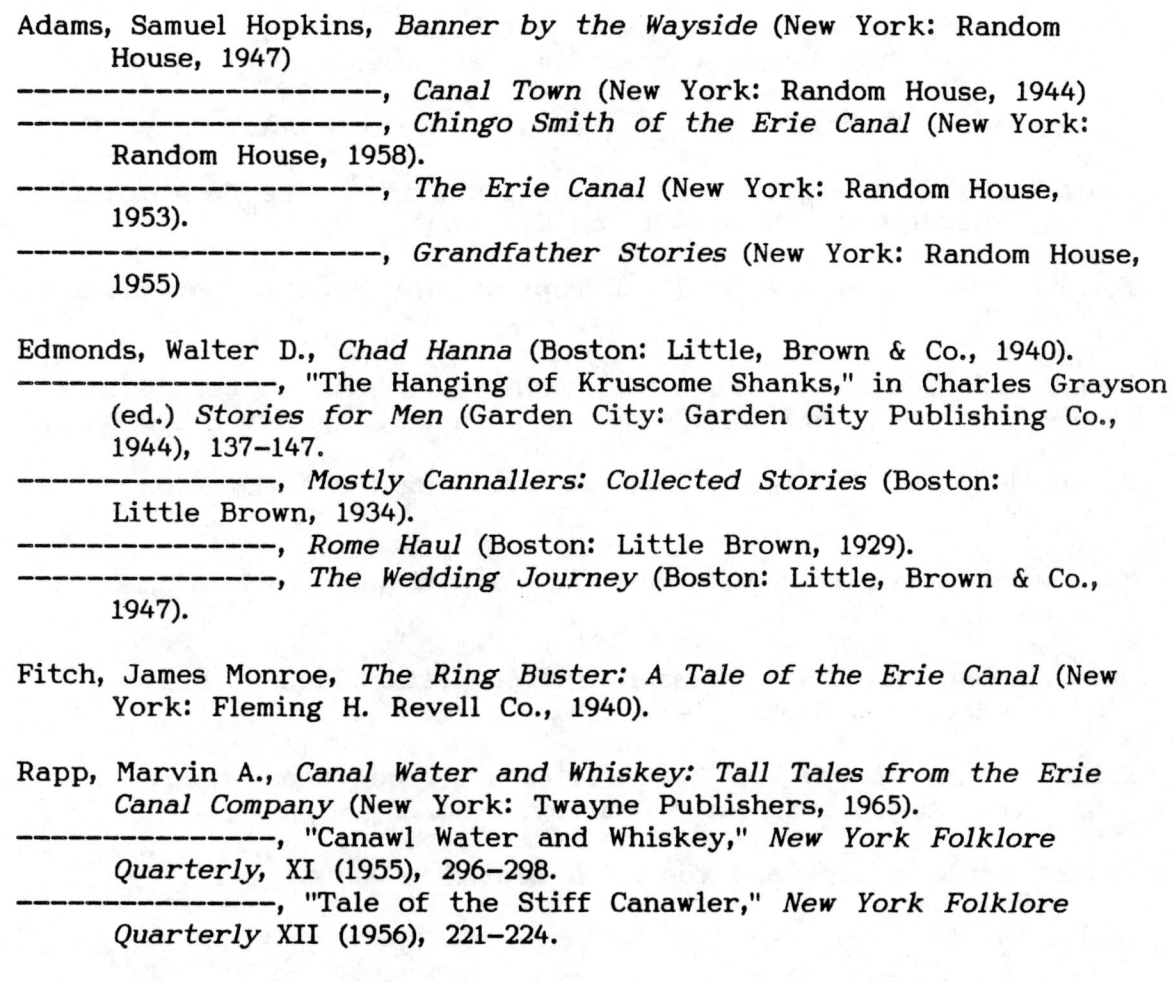

Adams, Samuel Hopkins, *Banner by the Wayside* (New York: Random House, 1947)
——————————, *Canal Town* (New York: Random House, 1944)
——————————, *Chingo Smith of the Erie Canal* (New York: Random House, 1958).
——————————, *The Erie Canal* (New York: Random House, 1953).
——————————, *Grandfather Stories* (New York: Random House, 1955)

Edmonds, Walter D., *Chad Hanna* (Boston: Little, Brown & Co., 1940).
——————————, "The Hanging of Kruscome Shanks," in Charles Grayson (ed.) *Stories for Men* (Garden City: Garden City Publishing Co., 1944), 137-147.
——————————, *Mostly Cannallers: Collected Stories* (Boston: Little Brown, 1934).
——————————, *Rome Haul* (Boston: Little Brown, 1929).
——————————, *The Wedding Journey* (Boston: Little, Brown & Co., 1947).

Fitch, James Monroe, *The Ring Buster: A Tale of the Erie Canal* (New York: Fleming H. Revell Co., 1940).

Rapp, Marvin A., *Canal Water and Whiskey: Tall Tales from the Erie Canal Company* (New York: Twayne Publishers, 1965).
——————————, "Canawl Water and Whiskey," *New York Folklore Quarterly*, XI (1955), 296-298.
——————————, "Tale of the Stiff Canawler," *New York Folklore Quarterly* XII (1956), 221-224.

White, Grace Miller, *From the Valley of the Missing* (New York: Grosset & Dunlap, 1911).

NEW YORK (cont.)

JUVENILE

Abbot, Jacob, *Marco Paul's Travels on the Erie Canal* (1843, reprint, Interlaken, N.Y: Empire State Books, 1987).

Berry, Erick, *Lock Her Through* (New York: Oxford Univ. Press, 1940).

Best, Herbert, *Watergate. A Story of the Irish on the Erie Canal* (Philadelphia: John C. Winston Co., 1951)

Edmonds, Walter, *Erie Water* (1933, reprint, New York: Random House, 1953).

Langdale, Hazel, *Jon of the Albany Belle* (E.P. Dutton & Co., 1943).
————————, *Mark of the Seneca Basin* (E.P. Dutton & Co., 1942).

Macdonald, Zillah K., *Two in Tow* (Boston: Houghton Mifflin Co., 1942),

Meadowcroft, Enid LaMonte, *Along the Towpath* (New York: Thomas Y. Crowell, 1940).

Orton, Helen Fuller, *The Treasure in the Little Trunk* (New York: Frederick A. Stokes, 1932).

Spier, Peter, (illustrator), *The Erie Canal*, (Young Reader's Press, Simon & Schuster Company, n.d.).

Tousey, Sanford, *Dick and the Canal Boat* (New York: Doubleday, 1943).

NEW YORK

Unpublished

Carp, Roger Evan, "The Erie Canal and the Liberal Challange to Classical Republicanism," PhD, Univ of North Carolina at Chapel Hill, 1986.

Cone, Gertrude E., "Studies in the Development of Transportation in the Champlain Valley to 1876," MA, Univ. of Vermont, 1945.

Ehrlich, Richard L., "The Development of Manufacturing in Selected Counties in the Erie Canal Corridor, 1815-1860," PhD, SUNY at Buffalo, 1972.

Larkin, F. Daniel, "The New York Years of John B. Jervis, Builder," PhD, SUNY, Albany, 1976.

Maddox, Vivian Dawn, "The Effect of the Erie Canal on Building and Planning in Syracuse, Palmyra, Rochester and Lockport, New York," 2 vols., PhD, Cornell University, 1976.

O'Hara, John E., "Erie's Junior Partner: The Economic and Social Effects of the Champlain Canal upon the Champlain Valley," PhD, Columbia, 1951.

Reineke, Robert, "Political Aspects of the Origin and Building of the Erie Canal," MA, Colgate Univ., 1951).

Siry, Steven Edwin, "De Witt Clinton and the American Political Economy: Sectionalism, Politics, and Republican Ideology, 1787-1828," PhD, Univ. of Cincinnati, 1986.

OHIO:

Bibliographical Aid

    Hasse, Adalaide R., *Index of Economic Materials in Documents of the States of the United States: Ohio, 1787-1904* 2 vols. (Washington: Carnegie Institution, 1912, Kraus reprint, 1965).

Other Works

    Ambler, Charles H, *History of Transportation in the Ohio Valley* (Glendale, California: Arthur H Clark Co., 1932).

    The Big Ditch, Small Stories of the Ohio Canals [Jim Baker's Historical Handbook Series], (Columbus Ohio: Ohio Historical Society, 2d ed. 1975), [cartoon history].

    Birch, Brian, "Taking the Breaks and Working the Boats: An English Family's Impressions of Ohio in the 1820s," *Ohio History*, XCV (1986), 101-118.

    Bogart, Ernest L., "Early Canal Traffic and Railroad Competition in Ohio," *Journal of Political Economy*, XXI (1913), 56-70.
    -----------------, *Internal Improvements and State Debt in Ohio* (New York: Longmans, Green and Co. 1924).

    Canal Fulton Heritage Society, *The Canal Era 1814-1913* (n.d.).

    Canals of Ohio, 1825-1913, (Columbus, Ohio: The Ohio Historical Society, 1971)

    Canal Society of Ohio, *The Ohio and Erie Canal from Lock 27 (Johnny Cake to Masillon* (Oberlin, Ohio: Canal Society of Ohio, 1977).

    "The Construction of the Ohio Canals." *Ohio Archaeological and Historical Quarterly*, XIII (1904), 460-481.

    "Commencement of the Ohio Canal at Licking Summit," *Ohio Archaeological and Historical Quarterly*, XXXIV (1925), 66-99.

    Dial, George W., "Construction of the Ohio Canals," *Ohio Archaeological and Historical Quarterly*, XIII (1904), 460-480.

    Doerschuk, A.N., "The Last Ohio Canal Boat," *Ohio Archaeological and Historical Society*, XXXIV (1926),   .

    Downes, Randolph C., *The Conquest and Canal Days* [Lucas County Historical Series, vols I and II], (reprint, Toledo, Ohio: Maumee Valley Historical Society, 1968).

OHIO (cont.)

>Downes, Randolph C. and Catheine G. Simonds, *The Maumee Valley USA, An American Story* (Toledo: Historical Society of Northern Ohio, 1955) Unit 5, Ch. I, 85-90.

>Droege, John, "Whitewater, A Canal Meant for Reconstruction," *CC*, No. 52 (Autumn 1980), 3.

>Droege, John, Barnett Golding and Richard Anderson, *The Ohio and Erie Canal from Lockborne to Caroll and Columbus Feeder Canal* (Akron, Ohio: Canal Society of Ohio, May 1974).

>"Electricity on the Miami and Erie Canal," *Scientific American*, XC (Jan 9, 1904), 25-26.

>Farrell, Richard T., Internal Improvement Projects in Southwestern Ohio," *Ohio History*, LXXX (Winter 1971), 4-23.

>French, Charles W., "Recollections of The Milan Canal," *Inland Seas*, I (October, 1945), 56.

>Frohman, Charles E., "The Milan Canal," *Inland Seas*, II (1946), 50-53.

>――――――――――, "The Milan Canal," *Ohio State Archaeological and Historical Quarterly*, LVII (1948), 237-246.

>Gamble, Jay Mark, "The Muskingum River," *National Waterways*, VIII (April 1930), 37-42; Part II, VII (May 1930), 49-51; Part III, VII (June 1930), 42-46, 63.

>――――――――――, *Steamboats on the Muskingum* (New York: Steamship Historical Society, 1971).

>Gard, R. Max and William H. Vodrey, Jr., *The Sandy and Beaver Canal* (East Liverpool, Ohio: East Liverpool Historical Society, 1952). [reprint in paper 1972].

>Geick, Jack, *A Photo Album of Ohio's Canal Era* (Kent, Ohio: Kent State University Press, 1987).

>George, John J., Jr., "The Miami Canal," *Ohio Archaeological and Historical Quarterly*, XXXVI (1927), 92-115.

>Guyton, Priscilla I., "John Hunt," *Northwest Ohio Quarterly*, LII (1980), 179-190, 214-226, 254-258.

>Hatcher, Harlan, *Lake Erie* (Indianapolis: Bobbs-Merrill Co., 1945, reprint, Greenwood Press, 1971), Chapter XI, "Canals Through the Barriers," 111-122.

OHIO (cont.)

Hirsch, Arthur, "The Construction of the Miami and Erie Canal," *Proceedings of the Mississippi Valley Historical Association* X (Nov 1921), 349-362.

Hull, Robert, *I Remember Roscoe. . .* (Bay Village, Ohio: Bob Hull Books, 1987).

Hunker, Robert L. (comp.), *The Cuyahoga Valley and the Ohio Canal* (Hudson, Ohio: n.p., 1974).

Huntington, C.C. and C.P. McClelland, *History of the Ohio Canals, Their Construction, Cost, Use and Partial Abandonment* (Columbus, Ohio: Ohio State Archaeological and Historical Society, 1905)

Kasper, Theobald W., "History of the Muskingum River Improvement," *AC,* No. 39 (November 1981), 6-7.

Ludwig, Charles, *Playmates of the Towpath: Happy Memories of Canal Swimmers' Society* (1929, reprinted, Cincinnati, Ohio: Ohio Book Store, 1986).

The Miami and Erie Canal: Symbol of an Era (Dayton, Ohio: Carillon Press, n.d.).

Morris, C. N., "Internal Improvements in Ohio," *Papers of American Historical Association,* III (1889), 351-379.

Nye, Pearl R., *Scenes and Songs of the Ohio-Erie Canal* (Columbus, Ohio: Ohio Historical Society, 1952, 1971).

Oda, James C., *Piqua and the Miami and Erie Canal* (Piqua, Ohio, Piqua historical Society, 1987)/

Ohio State Archaeological and Historical Society, *History of the Ohio Canals* (Columbus, Ohio, Fred J. Heer, 1905)

Porter, Burton, P., *Old Canal Days* (Columbus, Ohio: Heer Printing Company, 1942).

Ransom, Roger L., "Social Returns from Public Transport Investment: A Case Study of the Ohio Canal," *Journal of Political Economy,* LXXVIII (1970), 1041-1060.

Rodabaugh, James H. (ed.), "From England to Ohio, 1830-1832, 'The Journal of Thomas K. Wharton'," *Ohio Historical Quarterly,* LXV (1956), 1-27, 111-151. [contains canal sketches]

OHIO (cont.)

Ross, David F., "Living History: The Muskingum River," *AC*, No. 63 (November 1987), 10-12.

Scheiber, Harry N., "Alfred Kelley and the Ohio Business Elite, 1822-1859," *Ohio History*, LXXXVII (1978), 365-392.
――――――――――――, "The Commercial Bank of Lake Erie, 1831-1843," *Business History Review*, XL (1966), 57-65.
――――――――――――, "Entrepreneurship and Western Development: The Case of Micajah T. Williams," *Business History Review*, XXXVII (1963), 344-368.
――――――――――――, "Land Reform, Speculation, and Government Failure: The Administration of Ohio's State Canal Lands, 1836-60," *Prologue: The Journal of the National Archives*, VII (Summer, 1975), 85-98.
――――――――――――, *Ohio Canal Era: A Case Study of Government and the Economy*, 1820-1861 (Athens, Ohio: Ohio Univ., Press, 1969).
――――――――――――, "Ohio Canal Movement, 1820-1825," *Ohio Historical Quarterly*, LXIX (1960), 231-256.
――――――――――――, "Ohio's Transportation Revolution--Urban Dimensions, 1803-1870," in John Wunder (ed.), *Toward An Urban Ohio* (Columbus, 1977).
――――――――――――, "The Pennsylvania & Ohio Canal: Transport Innovation, Mixed Enterprise, and Urban Commerial Rivalry, 1823-1861," *Old Northwest*, VI (1980), 105-135.
――――――――――――, "Public Canal Finance and State Banking in Ohio, 1825-1837," *Indiana Magazine of History*, LXV (1969), 119-132.
――――――――――――, "The Rate-Making Power of the State in the Canal Era: A Case Study," *Political Science Quarterly*, LXXVII (1962), 397-414.
――――――――――――, "Social Returns from Public Transport Investment: A Case Study of the Ohio Canal," *Journal of Political Economy*, LXXVIII (1970), 1041-1060.
――――――――――――, "State Policy and the Public Domain: Ohio Canal Lands," *Journal of Economic History*, XXV (1965), 86-113.
――――――――――――, "Urban Rivalry and Internal Improvements in the Old Northwest, 1820-1860," *Ohio History*, LXXI (1962), 227-242, 289-292.

Schule, William, "Mike Kirwan's Big Ditch," *Reader's Digest*, XC (June 1967), 59-64 [Proposed Lake Erie-Ohio River Canal].

Still, John, "Ethan Allen Brown and Ohio's Canal System," *Ohio Archaeological and Historical Quarterly*, LXVI (Jan. 1957), 22-56.

Teagarden, Ernest M, "Builders of the Ohio Canal, 1825-1832," *Inland Seas*, XIX (1963), 94-103.

OHIO (cont.)

Treverrow, Frank, *Ohio's Canals* (Oberlin, Ohio: author, 1973)

Van Fossan, W. H., "Sandy and Beaver Canal," *Ohio State Archaeological and Historical Quarterly*, LV (1946), 165–177.

Verity, Vic, John Droege and Ralph Ramey, *The Miami Canal from Cincinnati to Dayton and Warren County Canal* (Oberlin, Ohio: Canal Society of Ohio, May 1977).

Unrau, Harlan and Nick Scrattish, *Historic Structure Report: Ohio and Erie Canal Cayahoga Valley National Recreational Area* (National Park Service, 1984).

White, Wallace B., "The Ghost Port of Milan and a Druid Moon," *Inland Seas*, VI (1950), 211–221; VII (1951), 21–28, 81–90.

Wilcox, Frank, *The Ohio Canals*, selected and edited by William A. McGill (Kent, Ohio: Kent State University Press, 1969).

Woods, Terry K., "Early Trade on Ohio's Western Canal, 1827–1840," *AC*, No. 18 (August 1976), 4, 5.
——————————, "The Life and Times of Pearl R. Nye: Balladeer, Historian, and Survivor of Ohio's Canal Era," *CCHT*, VII (1988), 47–72.
——————————, "The Search for Lock #5 [Walhonding Canal]," *AC*, No. 73 (May 1990), 8–9.
——————————, *Twenty Five Miles to Nowhere: The Story of the Walhonding Canal*, with Guide (Coshocton, Ohio: Roscoe Village Foundation, 1978).

Unpublished

Preston, Daniel, "Market and Mill Town: Hamilton, Ohio, 1795–1860," PhD, Univ. of Maryland, 1987.

Ransom, Roger L., "Government Investment in Canals: A Study of the Ohio Canal, 1825–1860," PhD, University of Washington, n.d.

Scheiber, Harry N., "Internal Improvements and Economic Change in Ohio, 1820–1860," PhD, Cornell Univ., 1962.

Still, John S., "The Life of Ethan Allen Brown, Governor of Ohio," PhD, Ohio State Univ., 1951.

ILLINOIS

Bibliographical Aid

Hasse, Adelaide R., *Index of Economic Material in Documents of the States of the United States: Illinois, 1809-1904* (Washington: Carnegie Institution, 1909, Kraus Reprint), "Canals," pps 124-146.

Other Works

Barr, Vernon F., "The Illinois Waterway," *Western Illinois Regional Studies*, VII (1984), 77-86.

Boswell, Lewis B., "The Hennepin Canal," *National Waterways* VII (Sept 1929, 37-63.

Cain, Louis P., "The Creation of Chicago's Sanitary District and Construction of the Sanitary and Ship Canal," *Chicago History*, VIII (1979), 98-110.

"The Chicago Drainage Canal," *Scientific American*, LXXII (June 15, 1895), 369-370.

Clayton, John, "How They Tinkered With a River," *Chicago History*, ns, I (1970-1971), 32-46.

Clemensen, A. Berle, *Illinois and Michigan Canal National Heritage Corridor, Illinois; Historical Inventory, History and Significance* (Denver, National Park Service, 1985).

Cozen, Michael P. and Kay J. Carr (eds.), *The Illinois & Michigan Canal National Heritage Corridor: A Guide to Its History and Sources* (Dekalb, Ill: Northern Illinois Press, 1988).

"Design for Controlling Works at the Head of the Chicago Drainage Canal," *Engineering News*, XLV (Jan 3, 1901), 2-3.

Elazar, Daniel J., "Gubernatorial Power and the Illinois and Michigan Canal: A Study of Political Development in the Nineteenth Century," *Journal of the Illinois State Historical Society*, LVIII (1965), 396-423.

Emmons, Francis A., "The Operating Machinery on the Illinois Waterway," *National Waterways*, X (Jan 1931), 51-2.

Fogerty, D., "Jewels of the Rust Belt [Illinois and Michigan Canal National Heritage Corridor]", *Sierra*, LXX (Sept/Oct 1985), 35-36 +.

Gies, Joseph, *Wonders of the Modern World* (New York: Thomas Y. Crowell Co., 1966), Chapter 6, "A City Solves its Worst Problem: The Chicago Sewerage System," 82-101.

ILLINOIS- (cont.)

Griffin, Donald W., "Recollections of the Hennepin Canal," *Western Illinois Regional Studies*, IV (1981), 50-76.

Hansen, Harry, *The Chicago* (New York and Toronto: Farrar & Rinehart, 1942), 127-151.

Howe, Walter A., *Documentary History of the Illinois and Michigan Canal* (Springfield: State of Illinois Department of Public Works and Buildings, 1956).

Hoyt, Homer, *One Hundred Years of Land Values in Chicago* (Chicago: Univ. of Chicago Press, 1933), Chapter I, "The Canal Land Boom, 1830-42," 3-44.

"The Illinois State Waterway for Barge Navigation," *Engineering News-Record*, LXXXV (Dec 2, 1920), 1095-1098 [plans and maps].

Johnston, Thomas T., "The Great Water Way to Connect Lake Michigan with the Mississippi River, and Its Influence on Floods in the Illinois River," *Journal of the Association of Engineering Societies*, VI (1886-1887), 182-199.

Kappe, Gale, "Following History's Trail," *Chicago*, XXXVI (April 1987), 99-104.

Krenkel, John H., *Illinois Internal Improvements, 1818-1848* (Cedar Rapids, Iowa: Torch Press, 1958).

Lamb, John K., "Canal Boats on the Illinois and Michigan Canal," *Journal of the Illinois State Historical Society*, LXXI (1978), 211-222.
---------------, *A Corridor in Time* (Romeoville, Ill: the author, History Department of Lewis University, 1986).
---------------, "Early Days on the Illinois and Michigan Canal," *Chicago History*, III (1974-1975), 168-176.
---------------, "The Illinois and Michigan Canal and Town Development in Northern Illinois," *CCHT*, III (1984), 3-12.
---------------, "The Kankakee Navigation," *AC*, No. 29 (May 1979), 4.
---------------, *William Gooding: Chief Engineer, I. and M. Canal* (Lockport, Illinois: Illinois Canal Society, 1982).

Larson, John W., *Those Army Engineers: A History of the Chicago District, U.S. Army Corps of Engineers* (Washington, Government Printing Office, 1980).

Naujoks, Herbert H., *The Chicago Water Diversion Controversy* (Milwaukee: Great Lakes Harbor Association, 1947).

ILLINOIS (Cont.)

Newton, Gerald A, John McFarland and Donald W. Griffin, "The Hennepin Canal: New Life for an Old Waterway," *Western Illinois Regional Studies*, VII (1984), 34-46.

Pease, Theodore C., *The Frontier State*, vol II of *The Centennial History of Illinois*, (Springfield, Ill: Illinois Centennial Commission, 1918) [Ch. X, "The Internal Improvement"; Ch. XI, "The Wreck of the Internal Improvement System, 1837-1842"; Ch. XVII, "The Internal Improvement System, The Solution."

Pierce, Bessie Louise, *A History of Chicago*, 2 vols. (New York: Albert A. Knopf, 1937-1949).

Putnam, James W., "Economic History of the Illinois and Michigan Canal," *Journal of Political Economy*, XVII (May-July 1909), 272-295, 337-353, 413-433.
————————, *The Illinois and Michigan Canal: a Study in Economic History* (Chicago: Univ. of Chicago Press, 1918).

Randolph, Isham, "The Sanitary District of Chicago, and the Chicago Drainage Canal: A review of 20 Years of Engineering Work," *Engineering News*, LXII (July 22, 1909), 90-94.

"The Report of a Board of Engineers on the Illinois Section of the Lakes-to-the-Gulf Waterway," *Engineering News*, LXV (March 2, 1911), 269-271.

Shank, William H., "How the Chicago River Flow was Reversed," *AC*, No. 73 (May 1990), 4-5, 6.

Stephens, George E., "The Illinois Waterway," *National Waterways*, VII (Nov 1929), 43-50, 78-79.

Stuve, Bernard, "The State's Internal Improvement Ventures of 1837-1838," *Publications of the Illinois State Historical Library*, VII (1902), 114-125.

Temple, Wayne C., *Lincoln's Connections with the Illinois & Michigan Canal, His Return from Congress in '48 and His Invention* (Springfield: Illinois Bell, 1986).

Vierling, Philip E., *Hiking the Illinois and Michigan Canal and Exploring its Environs*, vol. I. LaSalle to the Fox River [4 pamphlets], (Chicago: Illinois Country Outdoor Guides, 1986).

Yeater, Mary M., "The Hennepin Canal," *AC* No 19 (November 1976), 7; No. 20 (February 1977), 3, 4; No 21 (May 1977), 3, 7; No. 22 (August 1977), 6; No. 24 (February 1978), 5, 6; No. 25 (May 1978), 6-7; No. 16 (August 1978), 5, 6.

ILLINOIS (cont.)

Unpublished

Fleming, George J., Jr., "Canal at Chicago: A Study in Political and Social History," PhD, Catholic University of America, 1950.

Henderson, Richard R., "Illinois and Michigan Canal State Trail: Archival Inventory and Guide," MA, Sagamore State University, Springfield, Illinois).

Krenkel, John H., "Internal Improvements in Illinois, 1919-1848." PhD, Univ. of Illinois, 1939.

INDIANA

Benton, Elbert J., *The Wabash Trade Route in the Development of the Old Northwest* (Baltimore: Johns Hopkins Studies in Historical and Political Science, XXI 1903)

Cammack, Eleanore A., "Notes on Early Wabash Steamboating: Early Lafayette," *Indiana Magazine of History*, L (1954), 30-50.

Clark, George P. (ed.), "Through Indiana by Stagecoach & Canal Boat: The 1843 Travel Journal of Charles H. Titus," *Inidana Magazine of History* LXXXV (1989), 193-235.

Comstock, Howard P., "History of the Canals in Indiana," *Indiana Magazine of History*, VII (March 1911), 1-15.

Cottman, George S., "The Wabash and Erie Canal," *Indiana Magazine of History*, III (1907).
———————, "The Wabash and Its Valley, Part II," *Indiana Magazine of History*, I (1905), 123-127.

Duden, Margaret, "Internal Improvement in Indiana, 1816-1848," *Indiana Quarterly Magazine of History*, V (1909), 160-170.

Esarey, Logan, *History of Indiana from Its Exploration to 1850* (Indianapolis, Ind: W.K. Stewart Co., 1915).
———————, "Internal Improvements in Early Indiana," *Indiana Historical Society Publications*, V (1912), pps. 47-158.

Fatout, Paul, "Canal Agitation at Ohio Falls," *Indiana Magazine of History*, LVII (1961), 279-309.
———————, "Canalling in the Whitewater Valley," *Indiana Magazine of History*, LX (1964), 37-78.
———————, *Indiana Canals* (Lafayette, Ind: Purdue Univ. Press, 1972, 1987).

Fickle, James E., "The 'People' versus 'Progress' in the Old Northwest: Local Opposition to the Construction of the Wabash and Erie Canal," *Old Northwest*, VIII (Winter 1982-1983), 309-328.

Garman, Harry O., *Whitewater Canal Cambridge City to Ohio River: A Pioneer Transportation Facility in Indiana Constructed 1836 to 1847* (n.p., author, 1944,).

Hackett, Leola, "The Wabash and Erie Canal in Wabash County," *Indiana Magazine of History*, XXIV (1928), 295-305,

Holliday, Joseph E., "The Reservoir Regulators of the Canal Period," *Indiana Magazine of History*, XXV (1929), 92-100.

INDIANA (cont.)

Maldonado, Edwin, "Urban Growth During the Canal Era: The Case of Indiana," *Indiana Social Studies Quarterly*, XXXI (Winter 1978-1979), 20-39.

Mayhill, Dora T., *Old Wabash and Erie Canal in Carroll County* (Knightstown, Ind: Banner Publishing, Co., 1953).

McCord, Shirley S. (conp.), *Travel Accounts of Indiana 1679-1961, A Collection of Observations by Wayfaring Foreigners, Intinerants and Peripatetic Hoosiers* (Indianapolis, Ind., Historical Bureau, 1970), numerous canal items including "Canal Boat Trip," [197-201] from J. Richard Beste, *The Wabash: or Adventures of an English Gentleman's Family in the Interior of America*, 2 vols (London, 1855).

McDaniel, Dennis K., "Water Over Water: Hoosier Canal Culverts, 1832-1847," *Indiana Magazine of History*, LXXVIII (Dec. 1982), 296-322, reprint distributed with *AC*, No. 45 (May 1983).

Meek, Thomas, "A Plan to Save the Whitewater Canal," *AC,*Bulletin No. 57 (May 1986), 4-5.

Miller, James M., "The Richmond and Brookville Canal," *Indiana Magazine of History*, I (Fourth Quarter, 1905), 189-194.
———————, "The Whitewater Canal," *Indiana Magazine of History*, III (1907), 108-115.

Newcomer, Lee, "A History of the Indiana Internal Improvement Bonds," *Indiana Magazine of History*, XXXII (1936), 106-115.

Poinsatte, Charles R., *Fort Wayne During the Canal Era, 1828-1855* (Indianapolis, Ind: Indiana Historical Collections, XLVI 1969).

Reser, William J., "The Wabash and Erie Canal at Lafayette," *Indiana Magazine of History*, XXX (1934), 311-324.

Trout, William E., "The Indiana Central Canal ," *AC*, No.   (May 1979), 7; reprinted in *The Best from American Canals* (1980), 47.

Unpublished

Wehr, Paul H., "Samuel Hanna: Fur Trader to Railroad Magnate," PhD, Ball State Univ., 1968.

KENTUCKY

Bibliographical Aid

Hasse, Adelaide R., *Index of Economic Material in Documents of the States of the United States: Kentucky, 1792-1904* (Washington: Carnegie Institution, 1910, Kraus Reprint, 1965), "Canals, 109-110.

Other works

Dohan, Mary Helen, *Mr. Roosevelt's Steamboat* (New York: Dodd, Mead & Co., 1981), [Louisville rapids, 63-89].

Kreipke, Martha, "The Falls of the Ohio and the Development of the Ohio River Trade, 1810-1860," *Filson Club Historical Quarterly*, LIV (1980), 196-217.

"Ohio River Locks at Louisville," *Engineering News*, LXX (Dec 18, 1913), 1238-1244.

Sprague, Stuart Seely, "The Canal at the Falls of the Ohio and the Three Cornered Rivalry," *Register of the Kentucky Historical Society*, LXXII (1974), 38-54.
——————————, "The Louisville Canal: Key to Arron Burr's Western Trip of 1805," *Register of the Kentucky Historical Society*, LXXI (1973), 69-81.

Thomas, Samuel W., "The Falls of the Ohio River and Its Environs: The Journals of Increase Allen Lapham for 1827-1839," *Filson Club Historical Quarterly*, XLV (1971), 5-34, 199-226, 315-341, 381-403.

Trescott, Paul B, "The Louisville and Portland Canal Company, 1825-1874," *Mississippi Valley Historical Review*, XLIV (1958), 686-708.

Verhoeff, Mary, *The Kentucky River Navigation* [Filson Club Publications] (Louisville, Ky: John P. Morton & Co., 1917).

MICHIGAN

Barnett, Leroy, "Lac La Belle: Keweenaw's First Ship Canal,"
Michigan History, LXIX (1985), 40-46.

Brockel, Harry C.,"Historical Perspective of the Milwaukee River
[includes Milkaukee and Rock River Canal]," *Inland Seas*, XXV
1969), 105-122.

Chaput, Donald and Jim Schutze, "The Gladstone- AuTrain Ship Canal,"
Inland Seas, XXVII (1971), 206-278.

Dunbar, Willis F., "The Erie Canal and the Settlement of Michigan,"
Detroit Historical Society Bulletin, (Nov 1964).

Fisk, Howland, "The Clinton-Kalamazoo Canal," C, No. 43 (November,
1982?), 9-10.

Ireland, Irma Thompson, "The Northern Canal of Michigan," *Inland
Seas*, VII (1951), 158-166.

Keith, Hannah K., "An Historical Sketch of Internal Improvem,ents in
Michigan, 1836--1846," *Michigan Political Science Association
Publications*, IV (1900), 9-11, 24-35.

Parks, Robert J., *Democracy's Railroads: Public Enterprise in
Jacksonian Michigan* (Port Washington, N.Y./London: Kennikat
Press, 1972).
----------------,"Public Railroads in Michigan, 1825-1846," *Kansas
Quarterly*, II (1970, 18-23.

Sturm, John A., "The Clinton and Kalamazoo Canal--The Glory That Was."
Northwest Ohio Quarterly, LV (1983), 74-97.

VanMeer, Leo, "Clinton-Kalamazoo Canal," *Michigan History Magazine*,
XVI (1932), 225-231.

Unpublished

Parks, Robert J., "The Democracy's Railroads, Internal Improvements in
Michigan, 1825-1846," [vols. 1 and 2], PhD, Michigan State
University, 1967.

IOWA

Swisher, Jacob A., "The Des Moines River Improvement Project," *Iowa
Journal of History and Politics*, XXXV (April 1937), 3-21.

WISCONSIN

Gambill, Edward L., "Mason C. Darling and the Growth of Fond du Lac," *Wisconsin History*, XLV (1961), 84-94.

Kleist, Frederica Hart, "Boats on the Portage Canal," *AC*, No. (February 1983), 7.
————————————, "Fox River in the State of Wisconsin," *AC*, No. 63 (November 1987), 8-9.
————————————, *Portage Canal History since 1834*, (Portage, [Wisc] Canal Society, 1983).

Kleist, Frederica Hart and Herb O'Hanlon (compilers), "History of Wisconsin's Portage Canal," *AC*, No. 16 (February 1976), 5.

Lamboley, Kathryn, "Milwaukee and Rock River Canal Unlocked Little But Controversy," *Wisconsin Then and Now*, XXI (Sept 1974), 2-3, 7.

Mermin, Samuel, *The Fox-Wisconsin Rivers Improvement: An Historical Study in Legal Institutions and Political Economy, 1820-1861* (Madison, Univ. of Wisconsin Press, 1968).

Ross, David F., "Fox River Crisis," *AC*, No. 73 (May 1990), 3.

Smith, Alice E., "Fox River Valley Paintings," *Wisconsin Historical Magazine*, LI (1968), 139-154, (canal paintings from the 1850s reprinted in color.

Wenslaff, Ruth Ann, " The Fox-Wisconsin Waterway," *Wisconsin Then and Now*, XIII (1966), 1-3.

Unpublished

McCluggage, Robert W., "The Fox-Wisconsin Waterway, 1836-1872, Land Speculation, and Regional Rivalries, Politics and Private Enterprise," PhD, Univ of Wisconsin, 1954.

MASSACHUSETTS

Bibliographical Aid

Hasse, Adelaide R., *Index of Economic Material in Documents of the States of the United States: Massachusetts, 1789-1904* (Washington:Carnegie Institution, 1908, Kraus Reprint), "Canals", pps. 53-55.

Other Works

Ackerman, John H, "The Cape Cod Canal," *American History Illustrated*, I (1866), 32-39.

Adams, Charles Francis, "The Canal and Railroad Enterprise of Boston," in Justin Winsor (ed.), *The Memorial History of Boston*, IV, 111-150, (Osgood & Co. 1883).

Bell, Daniel W., "The South Hadley Canal, 1795-1847," *Historical Journal of Massachusetts*, XIII(2), (1985), 162-171.

Bockelman, Charles A., "Thomas Sheldon, Forgotten Resident of Westfield," *Historical Journal of Western Massachusetts*, VI (1977), 54-60.

Camposeo, James M., "The History of the Canal System Between New Haven and Northampton," *Historical Journal of Western Massachusetts*, VI (1977), 37-53.

"The Cape Cod Canal," *Engineering News*, LXIX (Jan 9, 1913), 47-50.

"The Cape Cod Canal," *Scientific American*, CIX (Sept 6, 1913), 184-185.

Clarke, Nancy Stetson, *The Old Middlesex Canal*, [reprint], Easton, Pa: Center for Canal History and Technology, 1974, 1987.

Eddy, Caleb, *Historical Sketch of the Middlesex Canal* (Boston, J.N. Dickinson, 1843).

Farson, Robert H., *The Cape Cod Canal* (Middletown, Conn: Wesleyan Univ. Press, 1977).

Franceschi, J.M., "Hampshire and Hampden Canal," *AC*, No. 35 (November 1980), 7.

Frizell, Joseph P., "On a Ship Channel Across Cape Cod," *Journal of the Franklin Institute*, XCI (1871), 386-391; XLII (1871), 41-47.

MASSACHUSETTS—(Cont.)

Handlin, Oscar and Mary, *Commonwealth: A Study of the Role of Government in the American Economy, Massachusetts, 1774-1861* (New York, New York Univ. Press, 1947, Harvard Univ. Press, 1969).

Lewis, Edward A., *The Blackstone Line: the Story of the Blackstone Canal Co. and the Providence & Worcester R.R.* (Seekonk, Mass., The Baggage Car, 1974?).

MacElwee, Roy S., "The Cape Cod Canal," Part One, *National Waterways* VII (Dec 1929), 11-16, 56; Part Two, VIII (Jan 1930), 57-65.

Malone, Patrick M., *Canals and Industry: Engineering in Lowell, 1821-1880* (Lowell, Mass: Lowell Museum, 1975, 1983).
————————, "Engineering and Industry in Lowell: 1821-1880," in Kemp, Emory L. and Theodore A. Sande (edws.), *Historic Preservation in Engineering Works* (American Society of Engineers, 1981), 134-158.

Miles, Henry A[dolphus], *Lowell As It Was and As It Is* (1845, reprinted Arno, 197?).

Miller, Jacob W., "The Cape Cod Canal," *Rngineering News*, LXIV (Dec 1910), 651.
————————, "The Cape Cod Canal," *National Geographic*, August 1914), 185-190,
————————, *Cape Cod and its Canal* (New York; Boston, Cape Cod and New York Canal Co., 1914).

Molloy, Peter M. (director and editor), *The Lower Merrimack River Valley: An Inventory of Historic Engineering and Industrial Sites,* revised (North Andover, Mass: Merrimack Valley Textile Museum, 1978)

Parsons, William B., "Cape Cod Canal," *Annals of American Academy of Political and Social Science*, XXXI (1908), 81-91.

Proctor, Thomas C., "The Middlesex Canal: Prototype for American Canal Building, *CCHT,* VII (1988), 124-174.

Reid, William J., *The Building of the Cape Cod Canal* (Privately printed, 1961).
————————, "The Military Value of the Cape Cod Canal," *United States Naval Institute Proceedings*, XCI (1965), 82-91.

Roberts, Christopher, *The Middlesex Canal, 1793-1860,* (Cambridge, Mass: Harvard University Press, 1938).

MASSACHUSETTS (cont.)

Skerrett, R.G., "A Bigger and Better Cape Cod Canal," *Scientific American*, CLIV (Jan 1936), 26-27.

Snow, Edward R., "America's First Canal," *AC*, No. 52 (Feb. 1985), 12 [excerpted from *Yankee* Magazine, March 1966].

Verplanck, E.F., "Excavating Methods and Equipment of the Cape Cod Canal," *Engineering News*, LXXI (Feb 19, 1914), 389-393.

Fiction:
Peters, Wayne R., *This Enchanted Land, Middlesex Village* (Lowell, Mass: author, 1984).

Unpublished

Reid, William J., "The Cape Cod Canal," PhD, Boston University, 1958.

CONNECTICUT

"Farmington Canal Remains," *AC,* No. 37 (May 1981), 9.

Harte, Charles, *Connecticut's Canals* (Hartford, Conn: author, 1938), reprinted from 54the Annual Report of the Connecticut Society of Engineers, inc..

Hayes, Lyman, "The Navigation of the Connecticut River," *Proceedings of the Vermont Historical Society for the Year 1915,1916,* 51-86.

Heinz, Bernard, "The Farmington Canal," *AC,* No. 48 (February, 1984), 7-8, reprinted and abridged from the *Connecticut Magazine.*

Hill, Evan, *The Connecticut River* (Middletown, Conn: Wesleyan University Press, 1972).

Jacobus, Melanchthon W., *The Connecticut River Steamboat Story* (Hartford, Connecticut Historical Society, 1956), [Chapter I].

Kistler, T.M., *The Rise of Railroads in the Connecticut River,* XXIII, nos. 1-4, Smith College Studies in History, 1938, Chapter I, 1-29.

Love, W. De Loss, "The Navigation of the Connecticut River," *Proceedings of the American Antiquarian Society,* ns, XV (1903), 385-444.

McLain, Guy A., "Steam Power on the Connecticut," *Historical Journal of Massachusetts,* XIV(2), (1986), 135-145.

Martin, Margaret E., *Merchants and Trade of the Connecticut River Valley, 1720-1820,* XXIV, nos. 1-4, Smith College Studies in History, 1939.

Milkofsky, Brenda," Three Centuries of Connecticut River Shipping," *Sea History,* XXXVI (Summer 1985), 12-16.

Sloane, Eric, "The Farmington Canal," *American Heritage,* IX (Feb 1958), 98-100.

RHODE ISLAND

Bibliographical Aid

Hasse, Adalaide R., *Index of Economic Material in Documents of the States of the United States: Rhode Island 1789-1904* (Washington: Carnegie Institution 1908, Kraus Reprint, 1965).

Other Works

Booth, Anne, *Historic Structure Report: Patucket Canal and Northern Canal Lock Structures* (Lowell, Mass:, U. S. Department of Interior, 1983).

NEW HAMPSHIRE

Bibliographical Aid

Hasse, Adelaide R., *Index of Economic Material in Documents of the States of the United States: New Hampshire 1784-1904* (Washington: Carnegie Institution, 1907, Kraus Reprint, 1965), "Public Works—Canals," 4 items, p. 53

Other Works

Grimes, Gordon F., "The Winnipiseogee Canal," *Historical New Hampshire,* XXIX (1974), 1-19.

McFarland, Henry, "Canals, Stage Lines and Taverns," in J.O. Lyford (ed.), *History of Concord, New Hampshire, from the Original Grant in Seventeen Hundred and Twenty-five to the Opening of the Twentieth Century* (Concord:, Rumford Press, 1903), II, 832-841..

Smith, Norman W., "A Mature Frontier: The New Hampshire Economy, 1790-1850," *Historical New Hampshire,* XXIV (1969), 3-19.

Waterman, W.R.," Locks and Canals at the White River Falls," *Historical New Hampshire,* XXII (Autumn 1967), 23-54.

VERMONT

Bibliographical Aid

Hasse, Adelaide, R., *Index of Economic Material in Documents of the States of the United States: Vermont 1789-1904* (Washington: Carnegie Institution, 1907, Kraus Reprint, 1965), "Public Works-Canals," 5 items, p. 59.

Unpublished

Amrheim, Joseph, "Burlington, Vermont: The Economic History of a Northern New England City," PhD, New York University, 1958.

MAINE

Bibliographical Aid

Hasse, Adelaide R., *Index of Economic Material in Documents of the States of the United States: Maine, 1820-1904* (Washington: Carnegie Institution, 1907, Kraus Reprint, 1965), "Public Works-Canals," 2 items, p. 63.

Other Works

Anderson, Hayden L. V., *Canals and Inland Waterways of Maine* (Portland, Me: Maine Historical Society, 1982?).
——————————, "Penobscot Waterways: Canals and Waterway Improvements on the Penobscot River, 1816-1921," *Maine Historical Quarterly*, XIX (Summer, 1979), 21-46.

Baker, William A., *A Maritime History of Bath, Maine and the Kennebec River Region*, 2 vols (Bath, Maine: Marine Research Society of Bath, 1973), [Peterson Canal, II, 147-148, 168-169].

Dole, Richard P., "Downeast Pioneer: The Bangor & Piscataquis Canal & Rail-Road Company," *Railroad History*, CLII (Spring 1985), 42-47.

Dole, Samuel T., "The Cumberland and Oxford Canal," *Maine Historical Society Collections* 2d ser., IX (1898), 264-271.

Milliken, Philip I., *Notes on the Cumberland and Oxford Canal & the Origin of the Canal Bank* [1821] (4th printing, Portland Maine: 1971).

Seeton, Jack, "The Cumberland and Oxford Canal in Maine." *CC*, No. 11 (Winter 1970), 4.

MARYLAND

Chesapeake and Delaware Canal

"AN ABSTRACT of sundry Papers and Proposals for improming the Inland Navigation of Pennsylvanai and Maryland, by opening a Communication between the Tide-Waters of Delaware and Susquehannah, or Chesapeake Bay; with a Scheme for an easy and short Land-Communication between the Waters of Susquehannah on Christiana-Creek, a Branch of Delaware; to which are annexed some Estimates of Expence, &c." *Transactions of the American Philosophical Society Held At Philadelphia for Promoting Useful Knowledge*, I (1771), 293-300, (reprint, New York: Kraus Reprint, 1966).

Brown, Earl I., "The Chesapeake and Delaware Canal," *Transactions, American Society of Civil Engineers*, XCV (1931), 716-765.

"Chesapeake and Delaware Canal Water Wheel for Raising Water," *Journal of the Franklin Institute*, LV (1853), 93-95.

Gray, Ralph D., "The Early History of the Chesapeake and Delaware Canal," *Delaware History*, VIII (1959), 223-229, 354-399; IX (1960), 66-98.

—————————, "'The Key to the Whole Situation'- The Chesapeake and Delaware Canal in the Civil War," *Maryland Historical Magazine*, LX (March 1965), 1-14.

—————————, *The National Waterway: A History of the Chesapeake and Delaware Canal*, 1769-1965 (Urbana, Ill: Univ. of Illinois Press, 1967; revised ed., 1989).

—————————, "Philadelphia and the Chesapeake and Delaware Canal," *PMHB*, LXXXIV (1960), 402-7.

Haupt, Louis M., "The Chesapeake and Delaware Canal," *Journal of the Franklin Institute*, CLXIII (1907), 81-107.

Ludwig, Edward J., III, *The Chesapeake and Delaware Canal: Gateway to Paradise* (Elkton, Md: Cecil county Bicentennial committee, 1979)

Stapleton, Darwin H., and Thomas C. Guider, "Transfer and Diffusion of British Technology: Benjamin Henry Latrobe and the Chesapeake and Delaware Canal," *Delaware History*, XVII (1976-1977), 127-138.

Weslager, Clinton A., *Delaware's Forgotten River: The Story of the Christina* (Wilmington, Delaware: Hambleton Company, Inc., 1947), Chapter 9, "Paths, Canals and Rails".

MARYLAND(cont.)

Chesapeake and Ohio Canal

> Atwood, Albert W., "Potomac, River of Destiny," *National Geographic*, LXXXVIII (July 1945), 33-70.
>
> Baird, W. David, "Violence Along the Chesapeake and Ohio Canal: 1839," *Maryland Historical Magazine*, LXVI (Summer, 1971), 121-134.
>
> Barnhart, Michael D., (compiler and preparer), *Chesapeake & Ohio Canal National Historical Park: Georgetown Boundaries & Agreements September-1984* (C & O NHP, 1984).
>
> Barron, Lee D., *The Chesapeake and Ohio Canal: "As It Is and As It Was"* (n.p: Graphics Design, 1973).
>
> Bauman, Chris, (preparer), *Widewater-- An Assessment for Historic Restoration- September 1984* (C & O NHP, 1984).
>
> Bearss, Edwin C., "1862 Brings Hard Times to the Chesapeake and Ohio Canal," *West Virginia History*, XXX (1969), 436-462.
> ----------------, "War Comes to the Chesapeake and Ohio Canal," *West Virginia History*, XXIX (1968), 153-177.
>
> *Chesapeake and Ohio Canal*, U. S. Govt., 22d Congress [Doc. No. 18], House of Representatives, 1st Session, Dec. 19, 1831.
>
> Clague, William, *A Collection of Maps of the Chesapeake & Ohio Canal* 5th edition (n.p: author, 1961, 1967, 1977.
>
> Clark, Ella E., and Thomas F. Hahn, (eds.), *Life on the Chesapeake and Ohio Canal, 1859* (York, Pa: American Canal and Transportation Center, 1975).
>
> Colton, Robert, *The C & O Canal in Photographs* (Mt. Deseret, Maine: Windswept House Publishers, 1986).
>
> Fradin, Morris, "Barging into the Past," *Travel*, LXXIX (Sept 1942), 8-10, 30.
>
> Franklin, William M., "The Tidewater End of the Chesapeake and Ohio Canal, *Maryland Historical Magazine*, LXXX (1986), 289-304.
>
> Frantz, William W., "Trip of the new 'Rudder Grange' Aug 23, 1894," *CC*, No. 46 (Spring 1979), 6-9.

MARYLAND (Cont.)

Chesapeake and Ohio Canal

Hahn, Thomas F., *The C & O Canal Boatmen, 1892-1924*, Shepherdstown, W.Va: American Canal and Transportation Center, 1982).

——————, *The C. & O Canal: An Illustratee History*, drawings by Diana Suttenfield-Abshire (Shepherdstown, W. Va: American Canal and Transportation Center, 1981)

——————, *Chesapeake and Ohio Canal, Old Picture Album*, (Shepherdstown, W. Va: American Canal & Transportation Center, 1976)

——————, *The Chesapeake & Ohio Canal: Pathway to the Nations Capital* (Shepherdstown, W. Va: American Canal and Transportation Center, 1984)[also, Metuchen, N.J: Scarecrow Press, Inc., 1984, rewrite and combination of several earlier works.

——————, "The Paw Paw Tunnell," *AC*, No. 3 (November 1909), 3, 7.

——————, *Towpath Guide to the C & O Canal, Georgetown (Tidelock to Cumberland* [combined edition], (Shepherdstown, W. Va: American Canal and Transportation Center, 1982?).

Harvey, Katherine A., "The Civil War and the Maryland Coal Trade," *Maryland Historical Magazine*, LXII (1967), 361-380.

Heine, Cornelius W., "The Chesapeake and Ohio Canal: Testimony to an Age Yet to Come," *Records of the Columbia Historical Society*, 1966-1968, 57-70; reprinted in Francis Coleman Rosenberger, *Records of the Columbia Historical Society of Washington, D.C., 1966-1986* (Wash. D.C: published by the Society, 1969).

Johnston, Jay, "Waterway to Washington: the C & O Canal," *National Geographic*, CXVII (1960), 419-439.

Kytle, Elizabeth, *Home on the Canal: An informal history of the Chesapeake and Ohio Canal, and recollections of eleven men and women who lived and worked on it* (Cabin John, Md: Seven Locks Press, 1983).

Larson, John L., "A Bridge, A Dam, A River: Liberty and Innovation in the Early Republic," *Journal of the Early Republic*, VII (1987), .

Lee. Ronald F., "Chesapeake and Ohio Canal," ecords of the Columbia Historical Society, XL-XLI (1940), 185-195.

MARYLAND (cont.)

Chesapeake and Ohio Canal (Cont.)

Mastrangelo, Mike, *The Community of Four Locks* (C & O NHP, 1987)

Morris, Elwood, "Sketch of the Tunnel now under construction upon the Cheaspeake and Ohio Canal, at the Paw-paw bend of the Potomac River," *Journal of the Franklin Institute*, XXVII (1839), 24-27.

Morris, Richard B., "Andrew Jackson, Strikebreaker," *American Historical Review*, LV (1949), 54-68.

Sanderlin, Walter S., "Arthur P. Gorman and the Chesapeake and Ohio Canal: An Example in the Rise of a Political Poss," *Journal of Southern History*, XIII (1947), 323-337.
--------------------, "The Conflict of Loyalties on the Chesapeake and Ohio Canal," *Maryland Historical Magazine*, XLII ((September 1947), 206-213.
--------------------, *The Great National Project: a History of the Chesapeake and Ohio Canal Company*, Johns Hopkins Studies in Historical and Social Sciences, LXIV, No.1, 1946, reprint, Arno, 1976.
--------------------, "Vicissitudes of the Chesapeake and Ohio Canal During the Civil War," *Journal of Southern History*, XI (1945), 51-67.

Stegmaier, Harry I., Jr., David M. Dean, Gordon E. Kershaw and John B. Wiseman, *Allegheny County, A History* (Parsons, W. Va: McClain Printing Co., 1976).

Ward, George. W., *Early Development of the Chesapeake and Ohio Canal Project*, Johns Hopkins Studies No. 17, (9) (10) (11) (Baltimore: 1899, Johnson reprint, 1973).

Wolfe, George W. "Hooper", *I Drove Mules on the C & O Canal*, (Williamsport, Maryland: author, 1969), Thomas Hahn, editor, 4 editions plus 2 supplements (Shephardstown, W.Va:, American Canal and Transportation Center, n.d.)

MARYLAND (Cont.)

Juvenile:

Fradin, Morris, *Hey-y-y Lock* (Cabin John, Md: See-and-Know-Press, c.1974)

Other

Formwalt, Lee W., "A Conversation Between Two Rivers: A Detate on the Location of the U.S. Capital in Maryland," *Maryland Historical Magazine*, LXXI (1976), 310-321.

James, Alfred R., "Sidelights on Founding of the Baltimore and Ohio Railroad," *Maryland Historical Magazine*, XLVIII (1953), 267-304, [Includes discussion of inclined planes on canals and railroads].

Kanarek, Harold, "The U.S. Army Corps of Engineers and Early Internal Improvements in Maryland," *Maryland Historical Magazine*, LXXII (1977), 99-109.

Sanderlin, Walter S., "The Maryland Canal Project-An Episode in the History of Maryland's Internal Improvement," *Maryland Historical Magazine*, XLI (March 1946), 51-65.

Scharf, J. Thomas, *The Chronicles of Baltimore* (Baltimore: Turnbull Brothers, 1874).

Unpublished

Egerton, Douglas Rogers, "Charles Fenton Mercer and the Foundations of Modern American Conservatism," 2 vols., PhD, Georgetown Univ., 1985.

Sanderlin, Walter S., "A History of the Chesapeake and Ohio Canal," PhD, Univ. of Maryland.

DELAWARE

Bibliographical Aid

Hasse, Adelaide R., *Index of Economic Material in Documents of the States of the United States: Delaware, 1789-1904* (Washington: Carnegie Institution, 1910), "Canals," pps. 40-41.

WASHINGTON, D.C.

> Formwalt, Lee W., "Benjamin Henry Latrobe and the Development of Transportation in the District of Columbia," *Records of the Columbia Historical Society*, L (1980), 36-66.
>
> Green, Constance McLaughlin, "The Jacksonian 'Revolution' in the District of Columbia," *Mississippi Valley Historical Review* XLV (1959), 591-605.
>
> Heine, Cornelius W., "The Washington City Canal," *Records of the Columbia Historical Society*, LIII-LVI (1953-1956, pub. 1959).
>
> Schell, Ernest H., "Timber Creek to Murder Bay: Failuree of the Washington Canal," *AC*, No. 25, May 1978), 4-5, 7.
>
> Skramsted, Harold, "The Georgetown Canal Incline," *Technology and Culture*, X (1969), 549-560.

VIRGINIA:

> Adams, Herbert Baxter, "Washington's Interest in the Potomac Company," *Johns Hopkins University Studies in Historical and Political Science* III (1885), 79-91.
>
> *Alexandria Canal* (Brochure Prepared by Department of Planning and Community Development & Alexandria Archaeology, Office of Historic Alexandria, n.d.)
>
> Art.I--Annual Report of the President and Directors of the Board of Public Works, to the General Assembly of Virginia in pursuance of an Act, entitled, An Act creating a Fund for Internal Improvement, Richmond; 1818 pp 78., *North American Review*, n.s. VIII (1818-1819), 1-26. [includes detailed history of hydraulic engineering.
>
> Bacon-Foster, Mrs. Corra, "Early Chapters in the Development of the Potomac Route to the West," *Records of the Columbia Historical Society*, XV (1912), 96-322; reprint, N.Y., Burt Franklin, 1971).
>
> Brown, Alexander C., "America's Greatest Eighteenth Cantury Engineering Achievement," *Virginia Cavalcade*, XII (Spring 1963), 40-46.
> ------------------, "The Canal Boat 'Governor McDowell,' Virginia's Pioneer Iron Steamer," *Virginia Magazine of History and Biography*, LXXIV (1966), 336-345.
> ------------------, "Colonial Williamsburg's Canal Scheme," *Virginia Magazine of History and Biography*, LXXXVI (1978), 26-32.
> ------------------, *The Dismal Swamp Canal* (Hilton Village, Va: 1945).
> ------------------, *The Dismal Swamp Canal* (Chesapeake, Va: Norfolk County Historical Society, 1970).

VIRGINIA- (Cont.)

Brown, Alexander C., "The Dismal Swamp Canal," *American Neptune*, V (1945), 203-222, 297-310; VI (1946), 51-70.

―――――――――――, "The Guard Locks at Great Bridge," *AC*, No. 14 (August 1975), 3-4 [abridged from article in Autumn 1974 issue of *Virginia Cavalcade*].

―――――――――――, *Juniper Waterway: A History of the Albemarle and Chesapeake Canal* (Charlottsville, Va: University Press of Virginia, 1981).

―――――――――――, "The Potowmack Canal Today," *Virginia Cavalcade*, XIII (Summer 1963), 42-46.

Davis, Captain John W., "Washington and the Patowmack Company," reprinted from the *Daughters of the American Revolution Magazine*, October 1929, *CC*, No. 69 (Winter 1985), 10-12.

Dent, Richard J., "On the Archaeology of Early Canals: Research on the Patowmack Canal in Great Falls, Virginia," *Historical Archaeology*, XX (1986), 50-62.

Dunaway, Wayland F., *History of the James River and Kanawha Company*, Columbia Univ. Studies in History, Economics and P:ublic Law, CIV, No, 2, 1922; reprint, AMS Press, 1969).

Finch, Marianne, *An Englishwoman's Experience in America* (London: Richard Bentley, 1853). [James River & Kanawha, pps. 309-315]

Garrett, W.E., "George Washington's Patowmack Canal," *National Geographic*, CLXXI (1987), 716-733.

Goodrich, Carter, "The Virginia System of Mixed Enterprise: a Study of the Planning of Internal Improvements," *Political Science Quarterly*, LXIV (1949), 355-387.

"Great Falls Canal and Locks," from *Civil Engineering-ASCE*, November 1972, *CC*, No. 69 (Winter 1985), 13-15.

Hahn, Thomas F., *George Washington's Canal at Great Falls, Virginia* (Shepherdstown, W. Va: American Canal and Transportation Center, 1976).

Handley, Harry E., "The James River and the Kanawha Canal," *West Virginia History*, XXV (1964), 92-101.

Hobbs, T. Gibson, "Early Canal Boats on the James River and Kanawha Canal," *AC*, No. 24 (February 1978), 3, 6; No. 26 (August 1978), 7; No. 28 (February 1979), 5..

―――――――――――, "James River and Kanawha Canal," *AC*, No. 36 (February 1981), 4-5; No. 37 (May 1981), 4-5; No. 38 (August 1981), 7-8.

―――――――――――, "Virginia Wooden Locks," *AC*, No. 27 (November 1978), 4.

VIRGINIA (cont.)

    Kirkwood, James J., *Waterway to the West*, (Published by Eastern National Park & Monument Association, 1963).

    Littlefield, Douglas R., "Eighteenth Century Plans to Clear the Potomac River: Technology, Expertise and Labor in a Developing Nation," *The Virginia Magazine of History and Biography*, XCIII (1985), 291-322.

    ——————————————, "Maryland Sectionalism and the Development of the Potomac Route to the West, 1768-1826," *Maryland Historian*, XIV (Fall-Winter, 1983), 31-52.

    ——————————————, "The Potomac Company: A Midadventure in Financing an Early American Internal Improvement Project," *Business History Review*, LVIII (1984), 562-585.

    Mitchell, Vivienne, "The Alexandria Canal," *AC*, No. 36 (August 1981), 4-5.

    Moseley, T.A.E., "The James River and Kanawha Company," *Nautical Research Journal*, XXIII (1977), 15-18.

    Murphy, Paul A., "The Kanawha Canal," *Historical Preservation*, XXIII (1971), 4-11.

    Pickell, John, *A New Chapter in the Early Life of Washington in Connection with the Narrative History of the Potomac Company*, New York: D. Appleton & Co., 1856) [reprint 1970]

    " A Trip on the JR & K Canal [1864]", *AC*, No. 48 (February 1984), 2-3.

    Trout, William E., III, "The Goose Creek and Little River Navigation," *Virginia Cavalcade*, XVI (Winter 1967), 31-34.

    ——————————————, *A Guide to the Works of the James River and Kanawha Company from the City of Richmond to the Ohio River* (Lexington, Va: Virginia Canals & Navigation Society, 1986, 2d ed., 1988).

    ——————————————, "Is the Junction Canal a Myth?" *AC*, No. 34 (August 1980), 5.

    ——————————————, *The James River Batteau Festival Trail: A Guide to the James River and Its Canal, From Lynchburg to Richmond*, 2d ed. (Lynchburg, Va: Moore's Country Store, 1989).

    Tulip, Frederick, *The River That Was Potomac* (Alexanderia, Va: author, n.d.).

<u>Unpublished</u>

    Rice, Philip Morrison, "The Virginia Board of Public Works, 1816-1842," MA, Univ. of North Carolina, 1947.

WEST VIRGINIA

Dean, William H., "Steamboat Whistles on the Coal," *West Virginia History*, XXXII (1971), 267-278.

Hale, J.P., "History of the Great Kanawha River Slack Water Improvement," *West Virginia Historical Magazine*, I (1901), 49-75.

Ross, David F., "Cruising the Great Kanawha," *AC*, No. 59 (Nov. 1986), 8-9.

SOUTH CAROLINA

Crowson, E. Thomas, "Building the Landsford Canal," *South Carolina History Illustrated*, I (1976), 18-22, 56-61.

Hollis, Daniel W., "Costly Delusion: Inland Navigation in the South Carolina Piedmont," *Proceedings of the South Carolina Historical Association*, 1968, 29-43.

Kohn, David and Bess Glen (eds.), *Internal Improvements in South Carolina, 1817-1828*, Washington, D.C: private printing, 1938).

Levine, Ida L., "A Letter From William Moultrie at Charleston to George Washington at Mount Vernon, April 7, 1786," *South Carolina Historical Magazine*, LXXXIII (1982-1983), 116-120. [refers to proposed Santee-Cooper Canal]

Porcher, F.A., *The History of the Santee Canal* (1875; reprint, Moncks Corner, S.C: Public Service Authority, 1950).

"Art. VII.--1. Report of the civil and military engineer of the State of South Carolina for the year 1819.
2. Plans and progress of internal improvement in South Carolina, with observations on the advantages resulting therefrom, to the Agricultural and Commercial interests of the State. Columbia, 1820.
3. Report of the Board of Public Works to the legislature of South Carolina for the year 1820," *North American Review*, n.s. IV (1821), 142-154.

Richardson, Lewis W., "The Canals of South Carolina," *AC*," No. 4 (February 1973), 6; No. 5 (May 1973), 4-5.

Rogers, George C., Jr., *Evolution of a Federalist: William Loughton Smith of Charleston (1758-1812)* (Columbia: Univ. of South Carolina Press, 1962).

SOUTH CAROLINA (Cont.)

>Savage, Henry, Jr., *River of the Carolinas: The Santee* (New York, Reinhart, 1956; Chapel Hill, Univ. of North Carolina Press, 1968).

GEORGIA

>Heath, Milton S., *Constructive Liberalism: The Role of the State in Economic Development in Georgia to 1860* (Cambridge, Harvard Univ. Press, 1954), [particularly Chapter X].
>————————, "Laissez Faire in Georgia, 1732-1860," *Journal of Economic History*, III, Supplement (1943), 78-100.

>McQueen, A.S. and Hamp Mizell, *History of Okefenokee Swamp* (1926, reprinted by Charlton County Historical Society, Folkston, Georgia, 1984).

>Richardson, L. W., "The Canals of Georgia," CC, No. 6 (August 1973), 6; No. 7 (November 1973), 5.

>Thurston, William N., "The Apalachicola-Chattahoochee-Flint River Transportation Route in the Nineteenth Century," *Georgia Historical Quarterly*, LVII (Summer 1973), 200-212.

>Trowell, C.T., *The Suwanee Canal Company in the Okefenokee Swamp*, Occasional Paper No. 9 (Douglas: South Georgia College, 1984)

>Trowell, C.T. and R. L. Islar, "Jackson's Folly: The Suwannee Canal Company in the Okefenokee Swamp," *Journal of Forest History*, XXVIII (1984), 187-195.

Unpublished

>Heath, Milton S., "Public Co-operation in Railroad Construction in the Southern United States to 1861," PhD Disseration, Harvard Univ., 1937).

>Ward, W. C., Jr., "Internal Improvement in Georgia, 1817-1843," Unpublished manuscript, Emory University, Atlanta, 1934.

ALABAMA

>Martin. William E., *Internal Improvements in Alabama*, The Johns Hopkins Studies in Historical and Political Science, XX, no. 4, (1902), 127-208.

>Richardson, L. W., "The First Alabama Canal," AC, No. 20 (February 1977), 5, 6.
>————————, "The Muscle Shoals Canals," AC, No. 23 (November 1977), 4-5.

TENNESSEE

Campbell, T.J., *The Upper Tennessee: Comprehending Desultary Records of River Operations in the Tennessee Valley. . .* (Chattanooga: Author, 1932)

Folmsbee, Stanley J., *Sectionalism and Internal Improvements in Tennesses* (Knoxville: East Tennessee Historical Society, 1939.
——————————, *Sectionalism and Internal Improvements in Tennessee, 1796-1845* (Philadelphia: Univ. of Pennsylvania Press. 1939).

A History of Navigation on the Tennessee River System (Washington, D. C: United States Government Printing Office, 1937

Johnson, Leland R., "Army Engineers on the Cumberland and Tennessee, 1824-1854," *Tennessee Historical Quarterly*, XXXI (Summer 1972), 5-17.
——————————, *Engineers on the Twin Rivers: A History of the Nashville District Corps of Engineers United States Army* (Nashville: U.S. Army Engineer District, 1978).
——————————, *A History of the Operations of the Corps of Engineers United States Army in the Cumberland and Tennessee River Valleys* (Nashville, Tenn: Vanderbuilt University, 1972).

Richardson, L.W., "The Hiwassee Canal," *AC*, No. 17 (May 1976), 3, 6.
——————————, "The Tennessee Canal," *AC*, No. 21 (May 1977), 5, 7.

TENNESSEE-TOMBIGBEE WATERWAY

"First 'Tow' Thru Tenn-Tom," *AC*, No 52 (February 1985), 1, 2.

Herndon, G. Melvin, "A 1796 Proposal for a Tennessee-Tombigbie Waterway," *Alabama Historical Quarterly*, XXXVII (1975), 176-182.

Patterson, Carolyn B., "The Tennessee-Tombigbee Waterway: Bounty or Boonddoggle," *National Geographic*, CLXIX (1986), 364-387.

Rumsey, Marian, "Deep in the Heart of Dixie," *Cruising World*, XII (Dec 1986), 80-83.

"Tenn-Tom Dedicated," *AC*, No 54 (August 1985), 1, 3-5.

TEXAS

Baughman, James P., "The Evolution of Rail-Water Systems of Transportation in the Gulf Southwest, 1836-1890," *Journal of Southern History*, XXXIV (1968), 357-381.

Crotty, Charles," The Houston Ship Channel," *Engineering News*, LXXII (July 23, 1914), 188-189.

Haupt, Lewis M., "The Galveston Harbor Problem," *Journal of the Franklin Institute*, CXXXII (1891), 275-287.

Kelley, Pat, *River of Lost Dreams: Navigation on the Rio Grand* (Lincoln: Univ. of Nebraska Press, 1986).

MacElwee, Roy S., "The Houston Ship Canal," *National Waterways*, VIII (March 1930), 11-17.

MacFadyen, J. Tevere, "Houston's Seaway Shouldn't Work, Yet Somehow it Does," *Smithsonian*, XVI (Oct 1985), 88-98.

Miller, Thomas Lloyd, *The Public Lands of Texas 1519-1970* (Norman: Univ. of Oklahoma Press. 1972). Chapter VI, "Land for Internal Improvemnts."

Sibley, Marilyn McA., *The Port of Houston: A History* (Austin and London: Univ. of Texas Press, 1968.

Woodward, Earl F., "Internal Improvements in Texas Under Governor Peter Hansborough Bell's Administration, 1849-1853," *Southwestern Historical Quarterly*, LXXVI (1972), 161-182.

NORTH CAROLINA

"Art. 2--Internal Improvements in North Carolina. . ." *North American Review*, XII (1821), 16-38 [particularly 22-33].

Gould, Alden W (ed.), "Reflections on the Roanoke Canal," *AC*, No. 16 (February 1976), 1-2.

Hinshaw, Clifford R., Jr., "North Carolina Canals before 1860," *North Carolina Historical Review*, XXV (January 1948), 1-56.

Weaver, Charles Clinton, *Internal Improvements in North Carolina Previous to 1860*, Johns Hopkins University Studies in Historical and Political Science, XXI, nos. 3-4, March-April 1903, reprint 1971)

LOUISIANA

Becnel, Thomas A, *The Barrow Family and the Barataria and Lafourche Canal: The Transportation Revolution in Louisiana, 1828-1925* (Baton Rouge: Louisiana State Univ. Press, 1989)

Bolding, Gary, "The New Orleans Seaway Movement," *Louisiana History*, X (1969), 49-60.

Dabney, Thomas Ewing, "New Orleans' Industrial Canal," *Scientific American*, CXXII (1920), 304, 314, 316.

Hatcher, G.E., "Cotton Cargoes Add to Ouachita Traffic [Louisiana and Arkansas]", *National Waterways*, XII (January 1932), 27-29, 48.

MacElwee, Roy S., "The Inner Harbor Navigation Canal--New Orleans, Louisiana," *National Waterways* X (Jan 1931), 18-24.

Millet, Donald J., "The Saga of Water Transportation into Southwest Louisiana to 1900," *Louisiana History*, XV (1974), 339-355.

FLORIDA

Bennett, Charles E. [Congressman, 2d Florida Congressional District], "Early History of the Cross-Florida Barge Canal," *Florida Historical Quarterly*, XLV (1966), 132-144.

Blake, Nelson Manfred, *Land into Water -- Water into Land: A History of Water Management in Florida* (Gainesville, Fla: University Presses of Florida, 1980).

Coachman, W.F., Jr., "A Brief on the Trans-Florida Canal," *National Waterways*, XII (Jan 1932), 14-17, 46.

Florida Ship Canal Authority, *Documentary History of the Florida Canal; ten year period, January, 1927 to June 1936* [74th Congress, Senate Doc. No. 275] (Washington, Government Printing Office, 1936).

Johnson, Lamar, *Beyond the Fourth Generation* (Gainesville, Univer. Presses of Florida, 1974).

Knotts, A.F., "Waterways for Florida," *National Waterways*, VIII (June 1930), 15-21.

Lowry, Sumter, "A Canal Across Florida," *The Review of Reviews*, LXXXIX (May 1934), 40, 55.
--------------, "A Florida Ship Canal," *The Review of Reviews*, LXXXVI (August 1932), 41-2

Mueller, Edward A., "Kissimmee Steamboating," *Tequesta*, XXVI (1966), 53-88.

Owens, Harry P., "Port of Apalachicola," *Florida Historical Quarterly*, XLVIII (1969-1970), 1-25.
--------------, "Sail and Steam Vessels Serving the Apalachicola-Chattahoochee [1828-1861]," *Alabama Historical Review*,", XXI (1968), 195-210.

Read, Henry H., *The Waterways of Florida Illustrated* (New York, Savannah and Jacksonville: Read Press, 1921).

Rogers, Benjamin F., "The Florida Ship Canal Project," *Florida Historical Quarterly*, XXXVI (1957), 14-23.

Shaffer, Tom and Jean, "The Quick Cut to the Gulf," [Okeechobee Waterway] *Cruising World* VII (Nov 1981), 78-81.

Sewell, J. Richard, "Cross-Florida Barge Canal, 1927-1968," *Florida Historical Quarterly*, XLVI (1968), 369-383.

FLORIDA (cont.)

Stoesen, Alexander R., "Claude Pepper and the Florida Canal Controversy, 1939-1943," *Florida Historical Quarterly*, L (1972), 235-251.

"Troubled Waters of the Everglades: For a Hundred Years the Construction of Canals and Dams has Upset the Ecology of the South Florida Region," *Natural History*, XCIII (1984), 46(12)-   .

Trout, William E. and Alden W. Gould, "Tour of Florida's Canals," *AC*, No.   (May 1982), 10-11.

U. S. Army Corps of Engineers in Florida, *Water Resource Development* (Atlanta, Ga., U.S. Army Engineer Division, South Atlantic, January 1979).

Whitman, Alice, "Transportation in Territorial Florida," *Florida Historical Quarterly*, XVII (July 1938), 25-53.

Will, Laurence E., "Digging the Cape Sable Canal," *Tequesta*, XIX (1959), 29-63.

Unpublished

Barber, Henry Eugene, "A History of The Florida Cross-State Canal," PhD, University of Georgia, 1969. [contains extensive bibliography]

GULF INTRACOASTAL WATERWAY

Alperin, Lynn M., *History of the Gulf Intracoastal Waterway* (n.p., U.S. Army Engineer Water Resources Support Center, January 1983).

Perry, George Sessions, "Now You Can Sail Through Texas," *Saturday Evening Post*, CCXXIII (July 15, 1950), 26, 131-134.

Texas Transportation Planning Division, *The Gulf Intracostal Waterway in Texas* (Austin: The Division, 1976).

Young, Gordon, "The Gulf's Workaday Waterway," *National Geographic*, CLIII (1978), 200-223.

"The Atlantic Coast Waterway," *Engineering News*, LXVII (May 16, 1912), 930-934.

Bailey, Anthony, "Our Footloose Correspondents Inside with the Coastal Queen," *New Yorker*, XL (Oct 31, 1964), 141-193.

Blanchard, Fessenden A., *A Cruising Guide to the Inland Waterway and Florida* (New York: Dodd, Mead & Co., 1958).

Brody, Catharine, "Florida Passage," *Saturday Evening Post*, CCX (Feb 26, 1938), 16-17, 45, 48, 52.

Fisher, Alan C., Jr. and photographs by James L. Amos, *America's Inland Waterway: Exploring the Atlantic Seaboard* (Washington, D.C: National Geographical Society, 1973).

Flagler, C.A.F., "Engineering Features of the Chesapeake and Delaware and Norfolk-Beaufort Waterway," *Annals of the American Academy of Political and Social Science*, XXXI (1908), 92-101.

Jarman, Rufus, "Enchanted Waterway," *Saturday Evening Post*, CCXXVII (Dec. 18, 1954), 22-23, 64-66.

Jones, Stuart E. and Dorothea, "Slow Boat to Florida," *National Geographic*, CXIII (1958), 1-68.

Luetscher, G.D., "Atlantic Coastwise Canals: Their History and Present Status," *Annals of the American Academy of Political and Social Science*, XXXI (1908), 73-80.

Miles, George F., "The Waterway of the Florida Coast Line Canal and Transportation Co.," *Engineering News*, LII (Aug 25, 1904), 163-165.

Spurr, Don, "The Boat Wore Blinders," *Cruising World*, XIV (May 1988), 80-85.

CALIFORNIA

Bibliographical Aid

Hasse, Adelaide R., *Index of Economic Material in Documents of the States of the United States, California, 1848-1904* (Washington: Carnegie Institution, 1908, Kraus Reprint), 4 items, p 59.

Other Works

Hardeman, N. F., "Overland in Cargo Ships: The Inland Seaport of Stockton, California," *Journal of the West*, XX (July 1981), 75-85.

Kemble, John H., *San Francisco Bay: A Pictorial Maritime History* (New York: Bonanza Books, 1957)

Thompson, John, "From Waterways to Roadways in the Sacramento Delta," *California History*, LIX (1980), 144-169.

Zelinsky, Edward Galland and Nancy Leigh Olmstead, "Upriver Boats—When Red Bluff Was the head of Navigation," *California History* LXIV (1985), 86-117, 161-162.

WASHINGTON [state] and OREGON

Ficken, Robert E, "'Seattle's Ditch': The Corps of Engineers and the Lake Washington Ship Canal," *Pacific Northwest Quarterly*, LXXVII (1986), 11-20.

Hardesty, W.P., "U. S. Improvements of the Columbia River, Oregon and Washington," *Engineering News*, LX (July 30, 1908), 109-117.

Harmon, Rick, "Alice Tomkins Fee: Growing Up on the Cascade Locks 'Reservation', An Interview," *Oregon Historical Quarterly*, LXXXVIII (1987), 285-307.

Hynding, Alan A., "Eugene Semple's Seattle Canal Scheme," *Pacific Northwest Quarterly*, LIX (1968), 77-87.

Kipp, R.H., "Columbia River--Second in America," *National Waterways*, IX (Aug 1930), 37-40.

MacElwee, Roy S., "The Lake Washington Ship Canal, Seattle, Washington," *National Waterways*, IX (Sept 1930), 21-28, 48-9, 64.

"Open River Number," *Oregon Historical Quarterly*, (June, 1915), 105-203, a special issue devoted to the completion and opening of the Dallas-Celilo Canal.

Purvis, N.H., "History of the Lake Washington Canal," *Washington Historical Quarterly*, XXV (1934), 114-127, 210-213.

Scott, Hugh A., "Reminiscence," *Oregon Historical Quarterly*, LXXXVIII (1987), 269-284.

Vincent, Fred W., "Connecting Idaho with the Sea," *Scientific American*, CXII (May 22, 1915), 476-477.

Willingham, William F., *Army Engineers and the Development of Oregon: A History of the Portland District U.S. Army Corps of Engineersa* (Portland, Ore.: Portland Distrit, Corps of Engineers, 1983).
————————————, "Engineering the Cascade Canal and Locks," *Oregon Historical Quarterly*, CXXXVIII (1987), 229-258.

Yound, Frederic G., "Columbia River Improvement and the Pacific Northwest," *Annals of the American Academy of Political and Social Science*, XXXI (1908), 189-202.

Baldwin, Leland D., "The Rivers in the Early Development of Western Pennsylvania," *WPHM*, XVI (1933), 79-98.

Brunot, Felix R., "Improvement of the Ohio River," *Journal of the Franklin Institute*, XCVII (1874), 305-327.

Carson, W. Wallacve, "Transportation and Traffic on the Ohio and the Mississippi Before the Steamboat," *Mississippi Valley Historical Review*, VII (1920-21), 26-38.

Cothell, Elmer L., *A History of the Jetties at the Mouth of the Mississippi River* (New York: J. Wiley & Son, 1881).

Covington, Samuel T., "Pioneer Transportation on the Ohio River," *Indiana Quarterly Magazine of History*, IV (1908), 12-20.

Cowdrey, Albert, *Lands End* [Corps History] (New Orleans: Corps of Engineers, 1977).

Dixon, Frank H., *A Traffic History of the Mississippi River System*, National Waterways Commission, Doc. 11, (Washington: Government Printing Office, 1909).

Dorsey, Florence L., *Master of the Mississippi: Henry Shreve and the Conquest of the Mississippi* (Boston: Houghton, 1941).
——————————, *Road to the Sea: The Story of James B. Eades and the Mississippi River* (New York: Rinehart & Co., 1947) [Mississippi-Ohio Waterway and the Techuantepec Ship-Railway].

Dobney, Frederick J., *River Engineers on the Middle Mississippi: A History of the St. Louis District U.S. Army Corps of Engineers* (Washington: Government Printing Office, 1978).

Eades, James B., "The Mississippi River Improvements," *Journal of the Franklin Institute*, CIV (1887), 209-216.

Ellet, Charles, Jr., "Contributions to the Physical Geography of the United States. Part I: Of the Physical Geography of the Mississippi Valley, with Suggestions for the Improvement of the Navigation of the Ohio and Other Rivers," *Smithsonian Contributions to Knowledge* II (1851), 1-58.
——————————, *An Essay on the Laws of Trade in Reference to the Works of Internal Improvement in the United States* (Richmond, Va: P.D. Bernard, 1839, reprinted, Augustus M Kelley, 1966).
——————————, *The Mississippi and Ohio Rivers* (Philadelphia: Lippencott, Granbo and Co., 1853).
——————————, *Report on the Improvement of the Kanawha and incidentally of the Ohio River by Means of Artificial Lakes* (Philadelphia: Collins, 1858).

MISSISSIPPI-OHIO CANALIZED RIVER SYSTEM, (cont.)

Feringe, P.A. and C. W. Schweizer, *One Hundred Years Improvement on the Lower Mississippi* (St. Louis: n.p., 1952).

Foreman, Grant, "River Navigation in the Early Southwest," *Mississippi Valley Historical Review*, XV (1928-29), 34-55.

Geyer, O.R., "Blasting a Canal Through a River Bottom," *Scientific American*, CXIV (May 6, 1916), 429.

Haites, Erik F. and James Mak, "Ohio and Mississippi River Transportation, 1810-1860," *Explorations in Economic History*, (Winter, 1970-71), 153-189.
──────────────────────────, "Steamboating on the Mississippi: A Study of a Purely Competitive Industry," *Business History Review*, XLV (Spring, 1971), 52-78.

Haites, Erik, James Mak and Gary M. Walton, *Western River Transportation: The Era of Early Internal Development, 1810-1860* (Baltimore: Johns Hopkins Univ. Press, 1975).

Hartley, C.W.S., "Sir Charles Hartley and the Mouths of the Mississippi," *Louisiana History*, XXIV (1983), 261-287.

Havighurst, Walter, *Voices on the River: The Story of the Mississippi Waterway* (New York: Macmillan, 1964).

Hawthorne, Lloyd, "Captain Henry Miller Shreve: Master of the Red," *North Louisiana Historical Association Journal*, II (1971), 1-6.

Henshaw, Leslie S., "Early Steamboat Travel on the Ohio River," *Ohio Archaeological and Historical Publications*, XX (1911), 378-402.

Hoagland, H.E., "Early Transportation on the Mississippi Before the Steamboat," *Journal of Political Economy*, XIX (1911), 111-123.

Hunter, Louis C., *Steamboats on the Western Rivers: An Economic and Technological History* (Cambridge: Harvard Univ. Press, 1949).
─────────────────, *Studies in the Economic History of the Ohio Valley*, XIX, Nos. 1 and 2, Smith College Studies in History, 1933-34; reprint, Johnson Reprint, 1970.

Jones, A.C., "On the Removal of Obstructions in the Mouth of the Mississippi," *Journal of the Franklin Institute*, XXXII (1841), 83-87.
─────────── , "Remarks on the Formation of Bars at the Mouth of the Mississippi River," *Journal of the Franklin Institute*, LVI (1853), 160-162.

MISSISSIPPI-OHIO CANALIZEAD RIVER SYSTEM (cont.)

Johnson, Leland H., *The Falls City Engineers: A History of the Louisville District Corps of Engineers United States Army* (Louisville: U.S. Army Corps of Engineers, 1975).

——————————, *The Falls City Engineers: A History of the Louisville District Corps of Engineers, United States Army 1970-1983* (Louisville: U. S. Army Engineer District, 1984).

Jones, Lawrence M., "The Improvement of the Missouri River and Its Usefulness as a Traffic Route," *Annals of the American Academy of Political and Social Science*, XXXI (1908), 178-188.

Kemper, James P. *Rebellious River* (Boston: Burce Humphries, Inc, 1949; Arno reprint, 1972).

Landon, Charles E., "Technological Progress in Transportation on the Mississippi River System," *The Journal of Business*, XXXIII (Jan. 1960, 45-62.

Lippencott, Isaac, "A History of River Improvement," *Journal of Political Economy*," XXII (1914), 630-666.

McCall, Edith, *Conquering the Rivers: Henry Miller Shreve and the Navigation of America's Inland Waterways* (Baton Rouge, La: Louisiana Univ. Press, 1984).

Merritt, Raymond H., *The Corps, The Environment, and the Upper Mississippi River Basin* Fort Belvoir: Historial Division, Office of the Chief of Engineers, 1984).

——————————, *Creativity, Conflict, and Controversy: A History of the St. Paul District, U.S. Army Corps of Engineers* (Washington: Government Printing Office, 1979).

Mills, Gary, *Of Men and Rivers* [Corps History] (Vicksburg: Corps of Engineers, 1978).

"The Mississippi Lock at Keokuk," *Engineering News*, LXX (Nov 13, 1913), 964-972.

Monette, John W., "The Progress of Navigation and Commerce on the Waters of the Mississippi River and the Great Lakes A.D. 1700 to 1846," *Publications of the Mississippi Historical Society*, VII (1903), 479-523.

MISSISSIPPI-OHIO CANALIZED RIVER SYSTEM (cont.)

Morris, Elwood, "On the Improvement of the Ohio River," *Journal of the Franklin Institute*, LXIII (1857), 1-15, 78-83, 145-153.

──────────, "On the Improvement of the Ohio River.--Review of the Practical Views of W. Milnor roberts, Esq, C.E.," *Journal of the Franklin Institute*, LXV (1858), 1-20; "Explanatory Remarks on the Review. . .," By W. Minor Roberts," *Journal of the Franklin Institute*, LXV (1858), 73-78.

"The Ohio River," from U.S. Army Corps of Engineers Navigation Chart Book, January 1976," *CC*, No. 50 (Spring 1980), 1-3.

Palmer, Charles K., "Ohio River Commerce, 1787-1936," *Indiana Magazine of History*, XXXIII (1937), 153-170.

"Present and Prospective Commerce of the Mississippi River from St. Louis to the Gulf of Mexico," [condensed from Appendix of Report of a Board of Engineers on the Project for a 14-ft Deep Waterway from the Lakes to the Gulf], *Engineering News*, LXII (Nov 4, 1909), 489-493.

Reuss, Martin, "The Army Corps of Engineers and Flood Control Politics on the Lower Mississippi," *Louisiana History*, XXIII (Spring 1982), 131-148.

──────────, *Army Engineers in the Memphis District* (Memphis: 1982).

──────────, "Andrew A. Humphreys and the Development of Hydraulic Engineering: Politics and Technology in the Army Corps of Engineers, 1850-1950," *Technology and Culture*, XXVI (1985), 1-33.

Roberts, W. Milnor, "Practical Views on the Proposed Improvement of the Ohio River," *Journal of the Franklin Institute*, LXIV (1857), 23-38, 73-85, 145-160, 217-231, 289-303, 261-372.

Sayenga, Donald, "The Ohio-Mississippi Waterway," *CCHT*, VII (1988), 73-123.

Sellers, Elizabeth M. "The Pittsburgh and Cincinnati Packet Line: Minute Book, 1851-1853," *WPHM*, XIX (1936), 243-254.

Stevens, George W., "Some Aspects of Early Intersectional Rivalry for the Commerce of the Upper Mississippi Valley," *Washington University Studies, Humanistic Series*, X (April, 1923), 277-300.

Vance, John L., "The Improvement of the Ohio River," *Annals of the American Academy of Political and Social Science*, XXXI (1908), 139-145.

MISSISSIPPI-OHIO CANALIZED RIVER SYSTEM (cont.)

    Way, R.B., "The Commerce of the Lower Mississippi in the Period 1830-1860," *Proceedings of the Mississippi Valley Historical Association*, X, extra number, (1920), 57-68.

    ----------, "Mississippi Improvement and Traffic Prospects," *Annals of the American Academy of Political and Social Science*, XXXI (1908), 146-163.

    ----------., "Mississippi Valley and Internal Mississippi Valley and Internal Improvements, 1825-40," *Mississippi Valley Historical Association Proceedings*, IV (1910-1911), 153-180.

    Wilby, Joseph, "A Victory of Man Over Nature [Eades]," *National Waterways*, VIII (June 1930), 9-14, 64; Part Two, IX (July 1930), 35-40, 64.

    Yager, R., *James Buchanan Eades: Master of the Great River* (Princeton, N.J: Van Nostrand, 1968).

Unpublished

    Haites, Erik F., "Ohio and Mississippi River Transportation, 1810-1860," PhD, Purdue University, 1969.

    Lowrey, Walter M., "Navigational Problems at the Mouth of the Mississippi River, 1698-1880," PhD, Vanderbuilt Univ., 1956.

    Palmer, Charles K., "Improvement and Navigation of the Ohio River, 1787 to 1925," MA, Indiana University, 1932.

Blee, C.E., "Development of the Tennessee River Waterway," *American Society of Civil Engineers Centennial Transactions*, 1953, 1132-1146.

Callahan, North, *TVA: Bridge over Troubled Waters* (Cranbury, N.J: A.S. Barnes and Co., 1980).

Clapp, Gordon R., *The TVA: An Approach to the Development of a Region* (Chicago:, Univ. of Chicago Press, 1955).

Creese, Walter L., *TVA's Public Planning: The Vision, The Reality* (Knoxville:, Univ of Tennessee Press, 1990)

Droze, Wilmon H, *High Dams and Slackwater: TVA Rebuilds a River* (Baton Rouge: Louisiana State Univ. Press, 1965).

Duffus, R.L., *Valley and Its People* (N.Y: Alfred A Knopf, 1949).

Hubbard, Preston J., *Origins of the TVA: The Muscle Shoals Controversy, 1920-1932* (Nashville, Tenn: Vanderbuilt Univ. Press, 1961).

King, Judson, *Conservation Fight from Theodore Roosevelt to the Tennessee Valley Authority* (Wash., D.C: Public Affairs Press, 1958).

Kyle, John H., *The Building of TVA; An Illustrated History* (Baton Rouge: Louisiana State Univ. Press, 1958).

Lilenthal, David, *TVA: Democracy on the March* (New York: Harper Bros., 1944, 1953).

Martin, Roscoe C. (ed.), *TVA: The First Twenty Years* (Knoxville: Univ. of Alabama and University of Tennessee Press, 1956).

Prichett, Charles H., *The Tennessee Valley Authority: A Study in Public Administration,* (Chapel Hill, N.C: Univ. of North Carlina Press, 1943).

Selznick, Philip, *TVA and the Grass Roots: A Study in the Sociology of Formal Organization* (Berkley and Los Angeles: Univ. of California Press, 1953; reprint with new preface, Harper & Row, 1966); reprint with new preface, N.Y: Harper & Row, 1966).

Tennessee River Navigation System: History, Development, and Operation [Technical Bulletin No. 25.] (Knoxville, Tenn: Tennessee Valley Authority, 1964).

SAULT STE. MARIE

Ballert, Albert G., "Commerce of the Sault Canals," *Economic Geograsphy*, XXXIII (1957), 135-162.

————————————, "The Soo Versus the Suez," *Canadian Geographical Journal*, LIII (Nov 1956), 160-167.

Baum, Arthur W., "World's Busiest Waterway," *Saturday Evening Post*, CCXXVII (June 4, 1955), 108-110.

Cohen, Nathan, "Battle for the Soo," *Reader's Digest*, XL (June 1942), 111-114.

deCotton, L., "A Frenchman Views Sault Ste Marie," translated with an introduction by George Joyaux, *Michigan History*, XXXVII (1953), 42-52.

Dickinson, John N., *To Build a Canal: Sault Ste. Marie, 1853-1854 and After*, (Columbus: Ohio State Univ. Press, 1981)

Fowle, Otto, *Sault Ste. Marie and Its Great Waterway* (New York: Putnam's Sons, 1925).

Harmon, R., "The Break at the Canadian Canal, Sault Ste. Marie, Ontario, 1909," *Inland Seas*, XXXV (1979), 104-109.

Havighurst, Walter, "The Way to the Big Sea Water," *American Heritage*, VI (1955), 20-25.

Holbrook, Stewart H., "The Saga of the Soo Canal," *American Mercury*, CX (1945), 450-454.

Jameson, Anne, "Impressions of Sault Ste. Marie, 1837," *Michigan History Magazine*, VIII (1924), 486-533.

Judson, Clara Ingram, *The Mighty Soo* (New York: Follet Publishing, 1955).

Lear, John, "Democracy Bets on the Soo," *Collier's*, CXXVII (Dec 2, 1950), 24-25, 57-59.

MacElwee, Roy S., "The St. Marys Falls Canal--'The Soo'," *National Waterways* IX (Oct 1930), 20-30, 62-64.

Mason, Philip, "The Operation of the Sault Canal, 1857," *Michigan History*, XXXIX (1955), 69-80.

Mayer, Harold M., "Great Lakes-Overseas: An Expanding Trade Route," *Economic Geography*, XXX (1954), 117-143.

SAULT STE. MARIE (Cont.)

    Moore, Charles, (ed. and comp.), *The Saint Marys Falls Canal: Exercises at the Semi-Centennial Celebration at Sault Sainte Marie, Michigan, August 2 and 3, 1905 together with a History of the Canal by John H. Goff, and Papers relating to the Great Lakes* (Detroit, Michigan: Semi-Centannial Commission, 1907).

    Neu, Irene, "The Building of the Sault Canal: 1852-1855," *Mississippi Valley Historical Review*, XV, (1953), 25-46.
    ----------, "The Mineral Lands of the St. Mary's Falls Ship Canal Company, " in David M. Ellis (ed.), *The Frontier in American Development: Essays in Honor of Paul Wallace Gates* (Ithaca and London, Cornell Univ. Press, 1969).

"New Canal and Locks at 'The Soo'," *Engineering News*, LXXI (March 5, 1914), 512-519, 879-886.

    Norton, Charles F., "Early Movements for the St Mary's Falls Ship Canal," *Michigan History*, XXXIX 1955), 257-280.

    Osborne, Brian S. and Donald Swainton, *The Sault Ste. Marie Canal: A Chapter in the History of Great Lakes Transport* (Ottawa: Parks Canada, 1986).

    Passfield, Robert W., *Technology in Transition: The 'Soo' Ship Canal 1889-1985* (Studies in Archaeology, Architecture and History, National Parks and Sites, Canadian Parks Service/Environment Canada, 1989).

    Rankin, Ernest H, "Canalside Superintendent," *Inland Seas*, XXI (1965), 103-114.

    Reynolds, Terry S., *Sault Ste. Marie: A Project Report* (Washington, D.C: Historic American Engineering Record, 1982).

"Sault Sainte Marie, Michigan," *Inland Seas*, XXIV (1968), 150-152, illus. 132-133.

    Wade, Herbert T., "Completing the World's Busiest Waterway!: The Fourth Lock at Sault Ste. Marie," *Scientific American*, CXVI (Feb 24, 1917), 202-203.

    Williams, Ralph D., *The Honorable Peter White: A Biographical Sketch of the Lake Superior Iron Country* (Cleveland: Penton Press Pub. Co., 1907).

"Work on the New Lock and Canal at Sault Sainte Martie, Mich.," *Engineering News*, LXVI (Sept 11, 1911), 275-279.

ST. LAWRENCE SEAWAY

Becker, William H. *From the Atlantic to the Great Lales: A History of the U.S. Army Corps of Engineers and the St. Lawrence Seaway* (Washington, D.C: Government Printing Office, 1984).

Brown, A.T., "New Saint Lawrence Seaway Opens the Great Lakes to the World," *National Geographic*, CXV (March 1959), 299-339.

Brown, George W., "The First St. Lawrence Deepening Scheme," *Michigan History Magazine*, X (1926), 593-605.

Burpee, L.H., "Canadian Section of the St. Lawrence Seaway," *Proceedings of the American Society of Civil Engineers*, (March 1960), paper no. 2420.

Chevrier, Lionel, *The St. Lawrence Seaway* (Toronto: Macmillan of Canada, 1959)
----------------, "The St. Lawrence Seaway and Power Project," *Geographical Journal*, CXIX (1953), 400-410.

Danielian, N.R., "The St. Lawrence Seaway," *Inland Seas*, VI (1950), 3-9.

Gallagher, James, "First U.S. Ore Carrier Trip into the Seaway," *Inland Seas*, XLI (1985), 89-99.

Gilmore, James, "The St. Lawrence Canals Vessel," *Transactions of the Society of Naval Architects and Marine Engineers*, 1956, 2-6, reprint distributed with Bulletin of American Canal Society, No 46 (February 1983).

Grothaus, W., and D. M. Ripley, "The St Lawrence Seaway, 27-ft Canals and Channels," *Proceedings of the American Society of Civil Engineers*, January 1958, paper no. 1518.

Hills, T.L., *The St Lawrence Seaway* (London: Methuen, 1959).

Hilton, Kenneth, "New York State's Response to St. Lawrence Seaway Support in the 1920s," *Inland Seas*, XXXVII (1981), 176-181; 252-257; XXXVIII (1982), 15-20.

Ireland, Tom, *The Great Lakes-St. Lawrence Deep Waterway to the Sea* (New York and London: G.P. Putnam's Sons, 1934.

Judson, Clara I., *St. Lawrence Seaway* (Chicago: Follett Publishing Company, 1959)

Legget, Robert F., *The Seaway* (Toronto: Clarke, Irwin & Co., 1974).

ST. LAWRENCE SEAWAY—(cont.)

LesStrang, Jacques, *The Great Lakes/St. Lawrence System* (Boyne City, Michigan: Harbor Home Publishers, 1985)

———————————, *Seaway: The Untold Story of North America's Fourth Seacoast* (Seattle, Washington: Superior Publishing Co., 1976).

Long, George W., B. Anthony Stewart and John E. Fletlcher, "Sea to Lakes on the St. Lawrence," *National Geographic*, XCVIII, (September 1950), 323-366.

Lower, A.R.M. (ed.), "Edward Gibbon Wakefield and the Beauharnois Canal: text of Memorandum," *Canadian Historical Review*, XIII (March 1932), 37-44.

Mabee, Carleton, *The Seaway Story* (N.Y: Macmillan, 1961).

MacElwee, Roy S., "Beauharnois Power and Navigation Canal," *Nations Waterways*, X (Feb 1931), 17-27, 57.

MacElwee, Roy S. and Alfred Ritter, *Economic Aspects of the Great Lakes-St. Lawrence Ship Canal* (N.Y: Ronald Press, 1921).

Moore, Edward T., "The Great Lakes—St. Lawrence Seaway and Power Project," *PDIA*, XX, # 5 (1952), 7-11.

Mouton, Harold G., Charles S. Morgan and Adak L. Lee, *The St. Lawrence Navigation and Power Prospect* (Wash., D. C: Brookings Institution, 1929).

Osborne, John, "The St. Lawrence Seaway," *Geographical Magazine*, XXXII (June and July 1959), 95-106, 139-146.

"The Seaway: Open for Business and Trouble, A portfolio of photographs by Erich Hartman," *Fortune*, LX (Sept 1959), 136-145.

Willowby, William R., *The Saint Lawrence Seaway: A Study in Politics and Diplomacy* (Madison: Univ. of Wisconsin Press, 1961).

Wright, C.P., *The St Lawrence Deep Waterway: A Canadian Appraisal*, (Toronto: Macmillian, 1935; reprint, Toronto, 1945).

CANADA:

Aiken, Hugh G. J., "Family Compact and the Welland Canal Company," *Canadian Journal of Economics*, XVIII (1952), 63-76.

────────────, *The Welland Canal Company* (Cambridge, Mass: Harvard University Press, 1954).

────────────, "Yates and McIntyre: Lottery Managers," *Journal of Economic History* XIII (1953), 36-57.

Andreas, Christopher, "The Canal at Sainte Marie Among the Hurons," *AC*, No 22 (May 1980), 3.

Angus, James T., *A Respectable Ditch: A History of the Trent-Severn Waterway, 1833-1920* (Downsville, Ont: McGill-Queens Univ. Press, available through Univ. of Toronto Press, 1988).

Armstrong, Frederick H., "John Armour of Dunnville: From Canal Supervisor to Village Patriarch," *Inland Seas*, XXIX (1973), 83-90.

Atwood, Margaret, "Summers on Canada's Rideau Canal," *Architectural Digest*, XLV (June 1988), 84+.

Auclair, Elie-J, "La Region de Soulanges et Son Canal," *Proceedings and Transactions of the Royal Society of Canada*, 3d series, XXXI, Section I (1937), 35-50.

Bain, J.W., "Surveys of a Water Route Between Lake Simcoe and the Ottowa River by the Royal Engineers 1819-1827," *Ontario History*, LI (1958), 15-27.

Ball, Norman R. (ed.), *Building Canada: A History of Public Works* (Toronto: Univ. of Toronto Press, 1988).

Beahen, William, *Development of the Severn River & Big Chute Lock Station* (Ottowa: Parks Canada, 1980)

Bergeron, Diane, Luc Bougie and Danielle Pineault (compilers), *Canal de Lachine, Atlas Historique* (Quebec: Parcs Canada, 1983).

Bleasdale, Ruth, "Class Conflicts on the Canals of Upper Canada in the 1840s," *Labour/LeTravailleur* [Canada], VII (1981), 9-39.

"The Break in the Cornwall, Ont. Canal and the Consequent Drawbridge Collapse," *Engineering News*, CX (July 9, 1908), 34-36 [illustrations].

Bush, Edward F., *Commercial Navigation on the Rideau Canal, 1832-1961*, History and Archeology Series, no. 54 (Ottawa: Parks Canada, 1981)

────────────, "When Steamboats Plied the Rideau River," *Canadian Geographic*, XCVIII (Feb/March 1979), 24-29.

CANADA (cont)

Cameron, Silver Donald, "Upgrading St. Peter's," *Canadian Geographic,* CVI (August/September 1986), 24-27.

Camu, Pierre, "The Traffic on the Upper St. Lawrence River, *Canadian Geographic,* III, nos. 1-4 1949).

A Canadian Enterprise--The Welland Canal: The "Merritt Day" Lectures-1078 - 1982 (St. Catharines, Ont: St. Catharines Historical Museum, 1984).

Canals of Canada (Department of Transport, Canada, 1953).

"The Chignecto Ship-Railway," *Engineering News and American Railway Journal,* XXII (July-December 1889), 218-219.

"Completion of the Surveys for the Montreal, Ottawa and Georgian Bay Ship Canal," *Engineering News,* LVIII (Oct 3, 1907), 370-371.

Coultee, C.R., "The Soulanges Canal Works, Canada," *Engineering News,* XLV (Apr 18, 1901), 274-278.

Curry, F.C., "The Rideau Canal System," *Inland Seas,* XXI (1965), 210-216.

Duquemin, Colin K. and Daniel J. Glenny, *Guide to the Grand River Canal* (St. Catharines, Ontario: St. Catharine Historical Museum, 1980).

Ecotour of the Rideau Canal (Smith Falls, Ontario: Parks Canada, 1978).

Espesset, Helene, "History of Quebec Canals: A Review of the Literature," *Research Bulletin,* Environment, Canada-Parks, Feb 1987, 1-17.

Elford, Jean, "The St. Clair River: Center Span of the Seaway," *Canadian Geographical Journal,* LXXXVI (1973), 18-23.

Francis, Daniel, *I Remember. . . An Oral History of the Trent-Severn Waterway* (Peterborough, Ont: Friends of the Trent-Severn Waterway, 1984).

Gay, Helen, "Down Stairs By Water to the Sea," *Travel,* LXV (July 1935), 28-32, 49-50.

Gillis, Sandra, "The Chambly Canal," *AC,* No. 15 (November 1975), 3-4.

CANADA (cont.)

Glazebrook, G.P. deT., *A History of Transportation in Canada* (Toronto: McClelland & Stewart, 1938), [reissued, 2 vols. paperback, 1964].

Grantmyre, Barbara, "The Canal That Bisected Nova Scotia," *Canadian Geographical Journal*, LXXXVIII (1974), 20-27.

Greenhill, Ralph, "The Peterborough Lift Lock," In Dianne Newell and Ralph Greenhill, *Survivals, Aspects of Industrial Archaeology in Ontario* (Erin, Ontario: The Boston Mills Press, 1989)

Greenwald, Michelle, Alan Levitt and Elaine Peebles, *The Welland Canals: Historical Resource Analysis and Preservation Alternatives* (Toronto: Ontario Ministry of Culture and Recreation, 1976, 1979).

Greening, W.E., "The Richelieu, Historic Waterway of Eastern Canada," *Canadian Geographical Journal*, LVI (March 1958), 84-93.

A Guide to the Grand River Canal (St. Catharine, Ontario: St. Catharine Historical Museum, 1984).

Haines, Charles, "Seaway Odyssey," *Canadian Geographia*, CV (Oct-Nov 1985), 8-23.

Harrington, Lyne, "Historic Rideau Canal," *Canadian Geographical Journal*, (December, 1947), 278-291.

Harrington, Lyne, "Historic Rideau Canal," *Canadian Geographical Journal*, XXXV (1947), 278-291.

Harrington, Lyne and Richard W., "The Welland Canal," *Canadian Geographical Journal*, XXXIV (1947), 202-215.

Heisler, John P., *The Canals of Canada* (Ottowa National Historical Sites Service, 1973).

Helleiner, F.M., "The Regionalization of a Waterway: A Study of Recreation and Boat Traffic," *The Canadian Geographer*, XXV (1981), 60-79.

Hill, B.E., "The Grand River Navigation Company and the Six Nations Indians," *Ontario History*, LXIII (1971), 31-40.

Hind, Edith J., "Troubles of a Canal-Builder: Lieut-Col. John By and the Burgess Accusations," *Ontario History*, LVII (1965), 141-147.

Jackson, John N., *Welland and the Welland Canal* (Belleville, Ontario: Mika Publishing, 1975).

CANADA–(Cont.)

Jackson, John N. and Fred A. Addis, *The Welland Canals: A Comprehensive Guide* (St. Catharines, Ont: Welland Canals Foundation, 1982).

Jarvis, Eric, "The Georgian Bay Ship Canal: A Study of the Second Canadian Canal Age: 1850-1915," *Ontario History*, LXIX (1977), 125-147.

Keating, E.H., "The Shubenacadie Canal," *Transactions of the American Society of Civil Engineers*, XII (1883), 436-440.

Keenan, W.E., "New 100-Ton Marine Railway on the Trent Severn Waterway," *AC*, No 32 (February 1980), 4.

Kendall, John C., "The Construction and Maintenance of the Coteau du Lac: The First Lock Canal in North America," *Journal of Transport History*, I (1971), 39-50.

Kingsford, William, *The Canadian Canals: Their History and Cost. . .* (Toronto: Rollo & Adams, 1865).

Lafreniere, Normand, *La Canalization du Saint-Laurent deux siecles de travaux 1779-1959* (Cahier No. 1, parc historique national Conteau-du-Lac, 1983).
—————————, *The Canalization of the St. Lawrence, 1779-1959* (Quebec: Parks Canada, 1983), [English translation .
—————————, *The Ottawa River Canal System* (Ottawa: Parks Canada, 1984.

Lanken, Dane, "Rediscovering Montreal's Lachine Canal," *Canadian Geographic*, CIII (April/May 1983), 46-51.

Legget, Robert F., *Canals of Canada* (Vancouver, B.C: Douglas, David & Charles, 1976)
—————————, *John By: Builder of the Rideau Canal* (Ottawa, Ontario: Historical Society of Ontario, 1982).
—————————, "The Jones Falls Dam on the Rideau Canal, Ontario, Canada," *Newcomen Society Transactions* [English], XXXI (1957-1958, 1958-1959), 205-218.
—————————, "The Ottawa River Canals and Portage Railways," *Newcomen Society Transactions* [English], XL (1967-1968), 61-73.
—————————, *The Ottawa River Canals and the Defence of British North America*, (Toronto: Univ. of Toronto Press, 1988).
—————————, *Ottowa Waterway: Gateway to a Continent* (Toronto: Univ. of Toronto Press).
—————————, *Rideau Waterway* (Toronto: Univ. of Toronto Press, 1955, 1972, 1986).

CANADA (cont.)

MacElwee, Roy S., "The Welland Ship Canal," *National Waterways*, IX (July 1930), 17-24; Part Two, IX (Aug 1930), 15-25, 62-64.

McLean, S.J., "The Georgian Bay Canal," *North American Review*, CXC (1909), 642-651.

McMillan, Ruth F., "The Welland Canal," *Inland Seas*, XXIII (1967), 316-319.

McNally, Larry S., *Water Power on the Lachine Canal, 1846-1900* (Quebec: Parcs Canada, 1982).

McNaughton, W.J., "Beauharnois: A Dream Come True, La Realization D'Un Reve," [in French and English], *Canadian Geographical Journal*, LXIV (Feb 1962), 40-65.

Mallory, Enid S., "The Trent-Severn Waterway in Ontario," *Canadian Canadian Geographical Journal*, LXVI (1963), 140-153.

Manning, Helen Taft, "E.G. Wakefield and the Beauharnois Canal," *Canadian Historical Review*, XLVIII (1967), 1-25.

Merritt, J.B., *Biography of the Hon. W.H. Merritt M.P.* (St. Catharines: 1887).

Moon, Robert (ed.), *Colonel By's Friends Stood Up* (Ottawa: Crocus House, 1979).

Morgan, R. J., "The Georgian Bay Canal," *Canadian Geographical Journal*, LXXVIII (March 1969), 90-97.

Naftel, William D., *The Rideau Waterway*, Occasional Publication, No 1. (Washington, D.C: Society for Industrial Archeology, April 1971).

Murphy, Rowley, " Memories of the Third Welland Canal," *Inland Seas*, XIX (1963), 260-265, illus. 298; XX (1964), 21-29, illus. 46-7; 120-128; 196-204, illus. 218; 295-302, illus. 309.

"The New Breakwater at Port Collborne, Ontario: Welland Canal Entrance," *Engineering News*, XLVII (May 15, 1902), 382-385.

"The New Welland Canal," *Scientific American*, CXXI (1919), 477, 500, 502.

"The New Welland Ship Canal," *Engineering News*, LXX (Sept 25, 1913), 598-602.

CANADA (cont.)

Newell, Dianne, "The Rideau Canal," in Newell and Greenhill, *Survivals*.

The Official Report on the Georgian Bay Ship Canal," *Engineering News*, LX (July 16, 1908), 69-70.

O'Neil, W.A., "The Welland Canal," *Proceedings of the American Society of Civil Engineers*, March 1958, paper # 1570,   .

Pammett, Howard, "The Steamboat Era on the Trent-Otonabee Waterway," *Ontario History*, LVI (1964), 67-103.

Parks Canada, *100 Ton Marine Railway Station 41--Big Chute* (Ottawa: 1977).

Passfield, Robert W., *Building the Rideau Canal, A Pictorial History* (Don Mills, Ont: Fitzhenry & Whiteside, 1982).
———————————, "Ordinance Supply Problems in the Canadas: The Quest for an Improved Military Transport System, 1814-1828," *HSTC Bulletin*, (Sept, 1981), 187-209.
———————————, "The Role of the Historian in Reconstructing Historic Engineering Structures: Parks Canada's Experience on the Rideau Canal, 1976-1983," *IA Journal of the Society for Industrial Archeology*, XI (1985), 1-28.
———————————, "Waterways," in Norman R. Ball (ed.), *Building Canada: History of Public Works* (Toronto, Buffalo and London, Univ. of Toronto Press, 1988).
———————————, "A Wilderness Survey: Laying Out the Rideau Canal, 1826-1832," *HSTC Bulletin, Journal of the History of Canadian Science, Technology & Medecine*, III (1983), 80-97.

Pentland, H.C., "The Lachine Strike of 1843," *Canadian Historical Review*, XXIX (Sept 1948), 255-277.

Petrie, Francis J, "First Welland Canal Opened Back in 1829," *Inland Seas*, XXV (1969), 62-63.

Raudzens, George, *The British Ordinance Department and Canada's Canals 1815-1855* (Waterloo, Ont: Wilfrid Laurier Univ. Press, 1979).
———————————, "The Military Impact on Canadian Canals, 1815-1825," *Canadian Historical Review*, LIV (1973), 273-286.

Richards, Terence, "Proposals for a Chatham Ship Canal, 1857-1893," *Inland Seas*, XXIV (1968).

Richardson, Boyce, "Steamers and Dreamers of the Upper Ottawa," *Beaver*, LXXVIII(3) (1988), 17-24.

CANADA (Cont.)

Sainsbury, G.V., "Re-routing the Historic Welland Canal," *Canadian Geographical Journal*, LXXXVIII (September, 1974), 36-43.

Sevigny, P.-Andre, "Le Commerce du ble' et la navigation dans le bas Richelieu avant 1849," *Review d'historie de l'Amerique francaise*, XXXVIII (Summer 1984), 5-21. [in French]

————————, *Trade and Navigation on the Chambly Canal: A Historical Overview* (Parks Canada, Canadian Gov't Printing Office, 1983); available in French as *Commerce et navigation sur le canal Chambly: apercu historique*.

————————, *The Work Force of the Richelieu River Canals, 1843-1950* (Ottawa: Parks Canada, 1983).

Shipley, Robert, *St. Catherines: Garden on the Canal* (Burlington, Ont: Windsor Publications, 1987).

Skidmore, P.G., "Canadian Canals to 1843," *Dalhousie Review*, LXI (1981-1982), 718-734.

Springer, J.F., "Canada's Great Ship Canal," *Scientific American*, CXXIV (1921), 504-505.

Squires, Roger, "The Trent-Severn Waterway," *AC*, No. 35 (November 1980), 3, 8.

Stead, Robert J.C., "Taming the St. Lawrence," *Canadian Geographical Journal*, LI (Nov 1955), 176-189.

Styran, Robert M. and Robert R. Taylor, "The Welland Canal: Creator of a Landscape," *Ontario History*, LXXII (1980), 210-229.

Styran, Robert M., *The Welland Canals The Growth of Mr. Merrit's Ditch* (Erin, Ontario: Boston Mills Press, 1989).

TenCate, Adrian G., *The Rideau: A Pictorial History of the Waterway* (Brockville, Ont: Becancourt, 1981).

Theberge, Clifford and Elaine, *The Trent-Severn Waterway: A Traveller's Companion* (Toronto and Sarasota: Samuel Stevens, 1978).

Tulloch, Judith, *The Rideau Canal: Defense, Transport, and Recreation*, History and Archeology Series, no. 50, (Hull, Que: Parks Canada, 1981).

Van Allen, W.H., "Canal Systems of Canada," *Canadian Geographical Journal* LXI (1960), 152-169.

CANADA (cont.)

Walton, Denver L., "Ontario's Rideau Canal," *CC*, No. 49 (Winter 1980), 2.

Warwick, Peter, D.A., "Three Shipbuilders from the Welland Canal," *Inland Seas*, No. 24 (1987), 50-59.

Welch, Edward, *Sights and Surveys* (Ottowa: Historical Society of Ottawa, 1979), [Rideau Canal].

Wells, Kenneth M., *Cruising the Trent-Severn Waterway* (Toronto: McClelland & Stewart, 1964).

Williams, Jack, *Merritt- Canadian Before His Time* (St. Catharines, Ont: Stonehouse Publications, 1985).

Willis, John, "Le Canal de Lachine Jusqu'en 1870: Origin et Function d'un Canal Hydraulique," *International Journal of History and Technology* (April 1986). [in French]

Wolf, Bill, "Wonderful Waterway [Trent-Severn Canal]," *Staurday Evening Post*, CCXXXI (Sept 6, 1958), 32-33, 75.

Wylie, William N.T., "Lockmaster on the Rideau: The Life of Peter Sweeney at Jones Falls, 1839-1850," *Ontario History*, LXXIII (1981), 112-126.
———————————, "Poverty, Distress and Disease: Labour and the Rideau Canal, 1826-1832," *Labour/Le Travailleur*, XI (1983), 7-29.

Yoder, C.P. "Bill", "1921 Cruise on the Trent Canal," *CC*, No. 60 (Autumn 1982), 3-6.

Young, Anna G., "A Canaller's Soliloquy," *Inland Seas*, XXIII (1967), 3-14.

Unpublished

Bleasdale, Ruth Elizabeth, "Unskilled Labourers on the Public Works of Canada, 1840-1880," PhD., University of Western Ontario, Canada, 1984.

Cureton, Edward A., "The Lachine Canal," These, McGill University, 1957.

Tulchinsky, Gerald J.J., "The Construction of the First Lachine Canal, 1815-1826," These, McGill University, 1960.

PANAMA CANAL

Bibliographical Aid

Bray, Wayne D. (compiler), *The Controversy Over a New Canal Treaty Between the United States and Panama: a Selected Annotated Bibliography of United States, Panamanian, Columbian, French and International Sources* (Washington: Library of Congress, 1976).

Morrison, H. A., *List of Books and Articles in Periodicals Relating to Interoceanic General Works (Nicarague; Panama, Darien Canal and Railway Routes and the Valley of Atrato; Tehuantepec and Honduras; Suez Canal)* prepared in Office of Superintendent of Documents (Washington: Gov't Printing office, 1900).

Library of Congress, Congressional Research Service, *Background Documents Relating to the Panama Canal* (Washington: U.S. Government Printing Office, 1977).

Sullivan [John T.] & Cromwell (comp.), *Compilation of executive documents and diplomatic correspondence relative to a trans-isthmian canal in Central America. With specific reference to the treaty of 1846 between the United States and New Granada (U.S. of Colombia) and the "Clayton Bulwer" treaty of 1850 between the United States and Great Britain*, 3 vols. (New York: The Evening Post Job Printing Office, 1905).

Other Works

Abbot, Henry L., "Present State of the Panama Canal," *Annals of the American Academy of Political and Social Science*, CCCI (1908), 12-35.

Abbot, Willis J., *Panama and the Canal in Picture and Prose* (London et al: Syndicate Publishing Company, 1914).

"The Adopted Plan for the Panama Canal," *Engineering News*, LV (Mar 1, 1906), 241-242.

Ameringer, Charles D., "Bunau-Varilla, Russia, and the Panama Canal," *Journal of Inter-American Studies and World Affairs*, XII (1970), 328-338.
──────────────────, "Ohio and the Panama Canal," *Ohio History*, LXXIV (1965), 3-12.
──────────────────, "The Panama Canal Lobby of Philippe Bunau-Varilla and William Nelson Cromwell," *American Historical Review*, LXVIII (1963), 346-363.
──────────────────, "Philippe Bunau-Varilla: New Light on the Panama Canal Treaty," *Hispanic American Historical Review* XLVI (Feb 1966), 28-52.

PANAMA CANAL (cont.)

Ammen, Daniel, "American Isthmian Canal Routes," *Journal of the Franklin Institute*, CXXVIII (1889), 409-439.

——————, "Mr. de Lesseps and His Canal," *North American Review*, CXXX (Feb. 1880), 130-146.

——————, "Recollections of the Panama Canal Congress," *North American Review*, CLVI (1893), 136-148.

Anguizola, Gustav A., "Negroes in the Building of the Panama Canal," *Phylon*, XXIX (1968), 351-359.

——————————, *Philippe Bunau-Varilla: The Man Behind the Panama Canal* (Chicago: Nelson-Hall, 1980).

"Annual Report of the Isthmian Canal Commission to the Secretary of War," *Engineering News*, LVIII (Nov 28, 1907), 367-373 [illustrations].

Augelli, John P., "The Panama Canal Era: The 'made-in-America' era comes to a close," *Focus*, XXXVI (Spring 1986), 20-29.

Avery, Ralph E., *America's Triumph at Panama: Panarama and Story of the Construction and Operation of the World's Giant Waterway from Ocean to Ocean*, edited by William C. Haskins (Chicago: L.W. Walter co., c.1913)

Barrow, Robert M., "The First Panama Canal Crisis, 1904," *Caribbean Studies*, V (4), (1960), 12-27.

Belohlavek, John M., "A Philadelphia and the Canal: The Charles Biddle Mission to Panama," *PMHB*, CIV (1980), 450-461.

Billard, Jules B., "Panama, Link Between Oceans and Continents," *National Geographic*, CXXXVII (March 1970), 402-440.

Bishop, Joseph Bucklin, *The Panama Gateway*, new and revised edition (NewYork: Scribner, 1915).

Bunau-Varilla, Philipe, *Panama: the Creation, Destruction and Resurrection* (London: Constable & Co., ltd., 1913).

Burr, William H., *Ancient and Modern Engineering and the Isthmian Canal* (New York: J. Wiley & Sons, 1902).

——————, "The Proposed Isthmian Ship-Canal," *Scribner's Magazine*, XXXI (Feb 1902), 145-169.

Calhoun, Crede Haskins, "Water Supply of the Panama Canal," *Scientific American* CXXV (1921), 157, 171.

PANAMA CANAL   (Cont.)

Campbell, John F., "'Captain John,' The Panama Canal's First Pilot," *American Neptune*, XXIV (1964), 186-207.

"Canal through the Isthmus of Panama," *Journal of the Franklin Institute*, II (1826), 252-253.

Chaloner, W.H., "The Birth of the Panama Canal, 1869-1914," *History Today* [English], IX (1959), 482-492.

Chester, Colby M., "The Panama Canal," *National Geographic*, XVI (1905), 445-467.

Chevalier, Michael, "An Extract from *An Historical and Geographical Examination of the Isthmus of Panama: The different directions it may be cut and the means employed*," by Chevalier [Paris, Gosselin], *Journal of the Franklin Institute*, XXXIII (1847), 304-310, 361-366.

Coker, William S., "The Panama Canal Tolls Controversy: A Different Perspective," *Journal of American History*, LV (1968), 555-564.

Compton, George C., "Through the Big Ditch," *Americas*, V (Aug 1953), 24-27.

Conniff, Michael L., *Black Labor on a White Canal* (Pittsburgh: Univ. of Pittsburgh Press, 1985).

Cornish, Vaughan, "The Condition and Prospects of the Panama Canal," *Royal Geographical Journal*, XLIV (1914), 189-203.

Corthell, Elmer, *Interoceanic Problem and Its Scientific Solution* (New York: 1885).

Crichfield, George W., "The Panama Canal from a Contractor's Standpoint," *North American Review*, CLXXX (1905), 74-87.

Crowell, Jackson, "The United States and A Central American Canal, 1869-1877," *Hispanic American Historicl Review*, XLIX (1969), 27-52.

deLesseps, Ferdinand, "The Interoceanic Canal," *North American Review*, CXXX (Jan. 1880), 1-15.
----------------------, "The Panama Canal," *North American Review*, CXXXI (July, 1880).

PANAMA CANAL (cont.)

DuVal, Miles Percy, *Cadiz to Cathay: the Story of the Long Struggle for a Waterway Across the American Isthmus* (Palo Alto, California: Stanford Univ. Press, 1940, 1947, [also Oxford Univ. Press, 1940]; Greenwood Press, 1968).

----------------, *And the Mountains Will Move* (Palo Alto, Calif: Stanford Univ. Press, 1947).

Ealy, Lawrence O., *Yanqui Politics and the Isthmian Canal* (University Park and London: Pennsylvania State University Press, 1971).

"Extracts from the Report of Napoleon Garella, an Engineer appointed by the French Government to survey the Isthmus of Panama. Translated for the Journal of the Franklin Institute by Persifor Frazer, Esq.," *Journal of the Franklin Institute*, XLII (1846), 18-24; 73-85; 145-154; 217-225.

Galvini, W.H., "Recollections of J.F. Stevens, his work on the Panama Canal and the Railroads and of Senator Mitchell," *Oregon Historical Quarterly* XLIV (1943), 313-326.

Gatell, Frank G., "The Canal In Retrospect-Some Panamanian and Columbian Views," *Americas*, XV (1958), 23-36.

Gillette, Cassius E., "The Panama Canal: Some Serious Objections to the Sea Level Plan," *Engineering News*, LIV (July 27, 1905), 81-84.

Good, Richard U., "A Trip to Panama and Darien," *National Geographic*, I (1888-1889), 301-315.

Gothals, George W., "The Building of the Panama Canal," *Scribner's Magazine*, LVII (1915), 265-282, 395-418, 531-548, 720-734.

----------------, "Extracts from the Annual Report of the Isthmian Canal Commission," *Engineering News*, LXVIII (Nov 12, 1912), 939-947.

----------------, "The Panama Canal," *National Geographic*, XXII (Feb. 1911), 148-211.

Greb, C. A., "Opening a New Frontier: San Francisco, Los Angles and the Pacific Canal, 1900-1914," *Pacific History*, XLVII (1978), 405-424.

Grenville, J.A.S., "Great Britain and the Isthmian Canal, 1898-1901," *American Historical Review*, LXI (1955), 48-69.

[Grosvenor, Gilbert H.], "Progress on the Panama Canal," *National Geographic*, XVI (1905), 467-475.

PANAMA CANAL (cont.).

Hains, Peter C., "The Labor Problem on the Panama Canal," *North American Review.* CLXXIX (1904), 42-54.

Hall, Wm. Ham., "The Conflict of Engineers over Plans for the Panama Canal," *Engineering News,* LIV, Part I (Nov 30, 1905), 572-575; Part II (Dec 7, 1905), 589-593; Part III (Dec 14, 1908), 616-619.

Hammond, Rolt and C.J. Lewis, *The Panama Canal* (Muller, 1966).

Heald, Jean Sadler, *Picturesque Panama, the Panama Railroad, the Panama Canal* (Chicago: C. Teich & Co., 1928).

Heinrichs, Waldo, "The Panama Canal in History, Policy and Caricature," [a Review Article], *Latin American Research Review,* XVI (1982), 247-261

Hill, Robert T., "The Panama Canal Route," *National Geographic,* VII (1896), 59-64.

Hogan, J.M., "Theodore Roosevelt and the Heroes of Panama," *Presidential Studies Quarterly,* XIX (Winter, 1989), 79-94.

"How the Locks of the Panama Canal are Operated," *Scientific American,* CX (March 7, 1914), 205, 214, 216.

Inkster, Tom H., "John Frank Stevens, American Engineer," *Pacific Northwest Quarterly,* LVI (1965), 82-85.

Johnson, Emory R., "The Isthmian Canal: Factors Affecting the Choice of Route," *Quarterly Journal of Economics,* XVI (1902), 514-536.
————————, "Panama Canal Revenues and Finances," *American Philosophical Society Proceedings* LXXXVII(2) (1943), 175-188.
————————, "The Panama Canal," *National Geographic,* X (1899), 311-316.

Johnson, Willis F. *Four Centuries of the Panama Canal. . .With maps and illustrations* (New York: H. Holt and Co., 1906).

Kaplan, E.S., "William Jennings Bryan and the Panama Canal Tolls Controversy," *Mid-America,* LVI (April 1974), 100-108.

Keller, Ulrich, *The Building of the Panama Canal in Historic Photographs* (New York: Dover, 1983)

Kemble, John H., *The Panama Route 1848-1869* (Berkley and Los Angeles: Univ. of California Press, 1943).

Klette, Immanuel J., *From Atlantic to Pacific; A New Interocean Canal* (New York: Published for Council on Foreign Relations by Harper and Row, 1967).

PANAMA CANAL (cont.)

Lancaster, H.C., "De Lesseps on Panama, Nicaragua and Tehuantepec," *American Philosophical Society Proceedings* XCIV(3) (1950), 258-9.

Lee, W. Storrs, *The Strength to Move a Mountain* (New York: G.P. Putnam's Sons, 1958).

LeFeber, Walter, *Panama Canal: The Crisis in Historical Perspective* (New York: Oxford Univ. Press, 1978, 1989).

Ludlow, William, "The Trans-Isthmian Canal Problem," *Harpers*, XCVI (1898), 837-848.

MacElwee, Roy S., "The Panama Canal," *National Waterways*, VIII (April 1930), 11-19, 62-64; Part Two, VIII (May 1930), 41-48, 60; Part Three, VIII (June 1930), 33-41.

McCullough, David G., "A Man, A Plan, A Canal, Panama," *American Heritage*, XXII (June 1971), 64-71, 100-1-3.
――――――――――――――――, *The Path Between Seas: The Creation of the Panama Canal 1870-1914* (New York: Simon & Shuster, 1977).

MacDonald, N.P., "Britain and an Atlantic-Pacific Canal," *History Today* [British], VII (1957), 676-684.

McDowell, "Panama Canal Today," *National Geographic*, CLIII, (1978), 278-294

McGinty, Brian, "'The Land Divided, a World United': The Panama Canal," *American History Illustrated*, XII(2) (1977), 10-19.

Mack, Gerstle, *The Land Divided: A History of the Panama Canal and Other Isthmian Canal Projects* (New York: Octogan, 1974, reprint of 1944 edition.

Maddox, Robert J., "How the Panama Canal Came About," *American History Illustrated*, III (Dec. 1968), 36-43.

Marden, Luis, "Panama, Bridge of the World," *National Geographic,*, LXXX (Nov 1941), 591-630.

Menocal, A.G., "Intrigues at the Paris Canal Conference," *North American Review*, CXXXVII (1879), 288-293.

Miller, Hugh G., *The Isthmian Highway: A Review of the Problems of the Caribbean* (New York: Macmillan, 1929).

PANAMA CANAL (Cont.)

Minor, Dwight C., *The Fight for the Panama Route* (1940, reprint New York: Octagon Books, 1971)

"The Monroe Doctrine and the Isthmian Canal," *North American Review*, CXXX (May 1880), 499-511.

Morison, George S., "The Panama Canal," *Engineering News*, XLIX (March 5, 1903), 219-224.

Naughton, William A., "The Rails that Linked the Oceans," *Americas*, XVII (Feb 1965), 11-17.

"Navigating Lights on the Panama Canal," *Scientific American*, CX (March 7, 1914), 200, 214.

Neary, Peter F., "The Panama Canal: Tolls Dispute, 1912-14," *Journal of Transport History*, VII (1966), 173-179.

Nimmo, Joseph, "The Proposed American Interoceanic Canal in its Commercial Aspects," *National Geographic*, X (1899), 297-310,. [comments on, X 363-4].

"On the Best Means of Establishing Commercial Intercourse between the Atlantic and Pacific Oceans," *Blackwood's Magazine*, LIV (1843), 658-671.

Paddelford, Norman J., "American Rights in the Panama Canal," *American Journal of International Law*, XXXIV (1940), 416-442.
----------, *Maritime Commerce and the Future of the Panama Canal* (Cambridge, Maryland: Cornell Maritime Press, 1975).
----------, "Neutrality, Beligerency, and the Panama Canal," *American Journal of International Law*, XXXV (1941), 55-89.
----------, "Panama Canal in Time of Peace," *American Journal of International Law*, XXXIV (1940), 601-637.
----------, *The Panama Canal in Peace and War* (New York: Macmillan, 1943).

Petras, Elizabeth McLean, *Jamacian Labor Migration: White Capital and Black Labor* (Boulder, Colo: Westview Press, 1988),

Penfield, Frederick C., "Suez and Panama," *North American Review*, CLXXX (1905), 817-828.

Pennell, Joseph, *Joseph Pennell's Pictures of the Panama Canal* (Philadelphia & London: Lippencott, 1912).

PANAMA CANAL (cont.)

Perez, Raul, "A Colombian View of the Panama Canal Question," *North American Review*, CLXXVII (1903), 63-68.

Perez-Venero, Alexander, *Before the Five Frontiers: Panama from 1821-1903* (New York: AMS Press, 1978).

Petras, Elizabeth McLean, *Jamacian Labor Migration: White Capital and Black Labor, 1850-1930* (Boulder: Westview Press, 1988)

"The Report of the Board of Consulting Engineers for the Panama Canal," including the minority report, *Engineering News*, LV (Feb 22, 1906), 202-210; (March 1, 1906), 234-240; "Extracts from the Minority Report," (June 7, 1906), 623-627.

Riley, Glenda, "Women on the Panama Trail to California, 1849-1869," *Pacific Historical Review*, LV (1986), 531-548.

Rockwell, Arthur E., "The Lumber Trade and the Panama Canal," *Economic History Review* [British], XXIV (1971), 445-462.

Roosevelt, Theodore, "President Roosevelt's Special Message on the Panama Canal," *Engineering News*, LVI (Dec 20, 1906), 663-666.

Rubinoff, Ira, "Mixing Oceans and Species," *Natural History*, LXXIV (Aug/Sept 1965), 69-72.

Rubio, Angel, "In the Wake of the Chagres: The River that Linked the Oceans," *Americas*, VI (Oct 1954), 16-19, 30-31.

Scheips, Paul G., "Gabriel Lafond and Ambrose W. Thompson: Neglected Isthmian Promoters," *Hispanic American Historical Review*, XXXVI (1956), 211-228.

Shouts, Theodore P., "What has been Accomplished by the United States Towards Building the Panama Canal," *National Geographic*, XVI (1905), 558-564.

Schowalter, Joseph, "Battling with the Panama Slides," *National Geographic*, XXIII (Feb. 1912), 195-205.
———————————, "The Panama Canal," *National Geographic*, XXV (Feb. 1914), 133-153.

Sibert, William L. and John F Stevens, *The Construction of the Panama Canal* (New York and London: D. Appleton and Co., 1915).

PANAMA CANAL (cont.)

Simon, Maaron J., *The Panama Affair* (New York: Charles Scribner's Sons, 1971).

Smith, Darrell H., *The Panama Canal: Its History, Activities, and Organization* (Baltimore: Brookings Institution, "Service Monographs of the United States Government,", No. 44, 1927; reprint, N.Y: AMS, 1974.

Stevens, John F., *An Engineer's Recollections (New York: McGraw Hill, 1936) reprinted from Engineering News Record between March 21 and Nov 21 1935.*
----------------, "Is a Second Canal Necessary?" *Foreign Affairs*, VIII (1930), 417-429.
----------------, "Report of the Chief Engineer of the Isthmian Canal Commission," *Engineering News*, LV (Jan 4, 1906), 11-14.

Suarez, Omar Jaen, "Across the Isthmus," *Americas*, XXXIX (May-June 1987), 28-35.

Taft, William H., "Statement of . . ., before the Committee on Inter Oceanic Canals of the United States Senate," *Engineering News*, LV (May 10, 1906), 513-517.

"Testimony of John F. Stevens, Chief Engineer of the Isthmian Canal Commission before the Senate Investigating Committee," *Engineering News*, LV (Feb 8, 1906), 140-145.

Venable, A.L., "John T. Morgan, Father of the Inter-oceanic Canal," *Southwestern Social Science Quarterly*, XIX (1939), 376-387.

Wallace, John F., "Report of the Chief Engineer of the Isthmian Canal Commission," *Engineering News*, LIII (Apr 20, 1905), 422-427.

Whitehead, R.H., "Facts About the Panama Canal," *Scientific American*, CXVI (Feb 10, 1917), 161-162.

Williams, Mary Wilhelmina, *Anglo-American Isthmian Diplomacy, 1815-1915* (Washington, D.C: American Historical Association, 1916, reprint, Russell & Russell, Inc., 1965).

Wright, Luke E., "Extracts from the Annual Report of the Isthmian Canal Commission," *Engineering News*, LX (Dec 3, 1908), 601-604 [illustrations].

Wyatt, V., "The Panama Railway," *Journal of the Franklin Institute*, LXVII (1859), 301-307.

PANAMA CANAL (cont.)

Unpublished

Griffin, Walt, "George W. Goethals and the Panama Canal," PhD, Univ. of Cincinnati, 1988).

Skinner, J.M., "The New Panama Canal Company, 1894-1908,", PhD, Kent State, 1974.

NICARAGUA

Allen, Cyril, "Felix Belly, Nicaraguan Canal Promoter," *Hispanic American Historical Review*, XXXVII (Feb 1957), 46-59.

Ammen, Daniel, "The Nicaragua Route to the Pacific," *North American Review*, CXXXI (Nov. 1880).

Bailey, Thomas A., "Interest in a Nicaragua Canal, 1903-1931," *Hispanic American Historical Review* XVI (1936), 2-28.

Berry, Charles R., "Travel on the Nicaragua Route in 1865: The Diary of John Green Berry, Jr.," *Journal of Transport History*, 3rd series, VIII (1987), 81-97.

Childs, O.W., "Survey of the Nicaragua Route for a Ship Canal," *Journal of the Franklin Institute*, LXXXIX (1870), 380-389; XC (1870), 39-43, 98-109, 166-174, 238-247, 325-330; XCI (1871), 25-29, 249-253, 363.

Clayton, Lawrence A., "John Tyler Morgan y el Canal de Nicaragua, 1897-1900," *Anario Estudios Centroamericanos [Costa Rica]*, IX (1983), 37-53 [in Spanish].
——————————————,"The Nicaragua Canal in the Nineteenth Century: Prelude to American Empire in the Caribbean," *Journal of Latin American Studies*, XIX (Nov. 1987), 323-352.

Davis, A.P., "Nicaragua and the Isthmian Routes," *National Geographic*, X (1899), 247-266.

Davis, Geo. W., "The Nicaragua Canal," *Journal of the Franklin Institute*, CXXXIV (1892), 1-20, 109-131 [map].

Folkman, David I., *The Nicaragua Route* (Salt Lake City: Univ. of Utah Press, 1972).

Grant, Ulysses S., "The Nicaragua Canal," *North American Review*, CXXXII (Feb. 1881), 107-116.

NICARAGUA

Greely, A.W., "The Present State of the Nicaragua Canal,". *National Geographic (1896), 73–76.*

Harvey, Charles T., "The Nicaragua Canal," *Cosmopolitan* X (1890), 676–684 [illustrated].

Hayes, C. Willard, "An Assumed Inconstancy in the Level of Lake Nicaragua: A Question of Permanence of the Nicaragua Canal," *National Geographic,* XI (1900), 156–161.

————————, "Physiography of the Nicaragua Canal Route," *National Geographic,* X (1899), 233–246.

Henry, Arnold K., "The Proposed Nicaragua Canal," *National Waterways,* VIII (May 1930), 11–16, 64.

Hill, Roscoe R., "The Nicaraguan Canal Idea to 1913," *Hispanic American Historical Review* XXVIII (1948), 197–211.

Keasbey, Lindley Miller, *The Nicaragua Canal and the Monroe Doctrine: A Political History of Isthmus Transit, with Special Reference to the Nicaragua Canal Project and the Attitude of the United States Government thereto* (N.Y: G.P. Putnam's Sons, 1896)

————————————, "The Nicaragua Canal and the Monroe Doctrine," *Annals of the American Academy of Political and Social Science,* V, VII (1896), No. 1, 1–31.

Kuhn, Joachim, "Napoleon III und der Nicaraguakanal," *Hist. Zeitschrift* [W. German], CCVI(2), (1969), 294–319.

Martin, Lawrence and Sylvia, "The World's Most Unbuilt Canal," *Antioch Review,* III (2) (1943), 262–270.

Miller, J.M., "The Advantages of the Nicaragua Route," *Annals of the American Academy of Political And Social Science,* VII (1896), 32–37.

"The Nicaragua Ship Canal (with map inserts)," *Engineering News and American Railway Journal,* XXII (July–December 1889), 247–250, insert opposite 252.

Noble, Alfred, "Some Engineering Features of the Nicaragua Canal," *Journal of the Western Society of Engineers,* III (1898), 771–778.

Peary, R.E., "Across Nicaragua with Transit and Machete," *National Geographic,* I (1888–1889), 315–335 [illustrations and maps].

Reed, T. B., "The Nicaragua Canal," *North American Review,* CLXVIII (1899), 552–562.

NICARAGUA (cont.)

    Rippy, J. Fred, "Justo Rufino Barrios and the Nicaragua Canal," *Hispanic Americn Historical Review*, XX (1940), 190-197.

    Scheips, P. J., "United States Commercial Pressures for a Nicaragua Canal," *Americas*, XX (1964), 333-358.

    Sherwood, G.W., "The Nicaragua Canal," *Journal of the Franklin Institute*, CXXXIX (1895), 425-438.

    Sultan, Dan I., "An Army Engineer Explores Nicaragua: Mapping a Route for a New Canal Through the Largest of Central American Republics," *Bational Geographic*, LXI (May 1932), 593-627.

    Taylor, H.C., "The Nicaragua Canal," *Journal of the Franklin Institute*, CXXVI (1889), 32-47; 81-95 [maps and plans].

    Wheeler, E.S., "The Topography of the Nicaragua Canal Route and the Plans and Estimated Cost of Constructing the Canal," *Engineering News*, XLIV (July 12, 1900), 21-25 [map opposite p. 26].

    Williams M.H., *San Juan River-Lake Nicaragua Waterway, 1502-1921* (Baton Rouge: Louisiana State Univ. Press, 1971).

    Williford, Miriam, "Utilitarian Design for the New World: Bentham's Plan for a Nicaraguan Canal," *The Americas*, XXVII(1), (1970), 75-85.

Unpublished

    Keasbey, Lindley M., "The Early Diplomatic History of the Nicaragua Canal," PhD., Columbia College [sic. University]. 1890.

TEHUANTEPEC [Mexico] SHIP RAILWAY

Covarrubias, Miguel, *Mexico South: The Isthmus of Tehuantepec* (N.Y: Alfred A. Knopf, 1967), "Appendix: The Epic of the Tehuantepec Railway," 163-173.

Corthell, Elmer L., *The Tecuantepec Ship Railway* (Philadelphia: Franklin Institute, 1885).
————————————, "The Tehuantepec Ship Railway," *National Geographic*, VII (1896), 64-72 [map].

Dallas, G. W., "The Isthmus of Tehuantepec," *Journal of the Franklin Institute*, XLIV (1847), 15-21.

Eades, James B., "The Isthmian Ship-Railway," *North American Review*, CXXXII (March, 1881), 223-238.
————————————, "Shall We have a Canal or a Railway?" *Scientific American Supplement*, IX, No. 221 (March 27, 1880, 3514-3515.

"The Interoceanic Ship Railway," *Scientific American*, LI, No. 26 (Dec. 27, 1884), 423, 428-431.

"Mr. Eads' Ship Railway for the Isthmus," *Scientific American*, LXIII, No. 20 (Nov. 13, 1880), 303, 306, 308-309.

"Shall We Have a Canal or a Ship Railway?" *Scientific American*, XLI, No.5 (Aug. 2, 1879), 63.

"Ship Railways for Isthmus Crossings," *Scientific American*, XLII, No. 13 (March 27, 1880).

"The Tehuantepec Ship Railway," *Journal of the Franklin Institute*, CXIX (1885), 456-489 [illustrations and maps].

SUEZ CANAL

Ballert, see listing under Sault Ste. Marie, p.

Baer, Werner, "The Promoting and the Financing of the Suez Canal," *Business History Review*, XXX (1956), 361-381.

Blumberg, Arnold, "An Early Project for a Suez Canal," *Mariner's Mirror*, LXVIII (1982), 317-322.

Bradshaw, Dan F., "Stephenson, DeLesseps, and the Suez Canal: An Englishman's Blindspot," *Journal of Transport History*, IV (1978), 239-243.

Burchell, S.C., *Building the Suez Canal* (New York: Harper and Row, 1966).

Cameron, Rondo, *France and the Economic Development of Europe, 1800-1914: Conquests of Peace and Seeds of War* (Princeton, N. J: Princeton Univ. Press, 1961), "Suez," 472-481.

"Canal Across the Isthmus of Suez," *Journal of the Franklin Institute*, XLIV (1847), 79-81.

"The Canal Through the Isthmus of Suez," *Journal of the Franklin Institute*, LXI (1856), 150 with Plate #2

Danziger, C., "The First Suez Crisis," *History Today*, XXXII (Sept 1982), 3-7.

DeLesseps, *Lettres jounal et documents pour servir a l'histore du canal de Suez. . .* (Paris: Didier et Cie, 1875-1881).
---------, *The Suez Canal: Letters and Documents Descriptive of Its Rise and Progress in 1854-1855*, translated by N D'Anvers (London: Henry S. King and Co., 1876, reprinted, Scholarly Resources, Inc., 1976).

Duff, R.E.B., *100 Years of the Suez Canal* (London: Clifton Books, 1969).

Edgerton, Glen E., "An Engineer's View of the Suez Canal," *National Geographic*, CXI (1957), 123-140.

"Egypt and the Story of the Suez Canal," *Blackwood's Magazine* CVI (1867), 730-745

Fitzgerald, Percy, *The Great Canal of Suez*, 2 vols (London: Tinsley Bros., 1876).

SUEZ CANAL (cont.)

Graves, William, "New Life for the Troubled Suez Canal," *National Geographic*, CXLVII (1975), 792-817.

Hallberg, Charles W., *The Suez Canal* (New York: Columbia Univ. Press, 1931)

Hanson, B. and K. Tourk, "The Profitability of the Suez Canal as a Private Enterprise, 1859-1956," *Journal of Economic History*, XXXVIII (1978), 938-958.

Kaskey, Elizabeth, *Empress Eugenie's State Visit to the Opening of the Suez Canal, 18969: an Album of drawings by Alfred-Henri Darjou and P. Montand* (New York: Shepherd gallery, 1988).

Kinross, Lord, *Between Two Seas: The Creation of the Suez Canal* (New York: William Morrow & Co., 1969).

Landes, David S., *Bankers and Pashas: International Finance and Economic Imperialism in Egypt* (Cambridge, Mass: Harvard Univ. Press, 1958).

MacElwee, Roy S., "The Suez Canal," *National Waterways*, VI (Jan 1929), 14-24, 62-64.

Marlowe, John, *World Ditch: The Making of the Suez Canal* (New York: Macmillan, c. 1964)

Mitchell, Henry, "The Coast of Egypt and the Suez Canal," *North American Review*, CIX (1869), 476-509.

Mountjoy, Alan B., "The Suez Canal at Mid-Century," *Economic Geography*, XXIV (1958), 155-167.

Moore, W. Robert, "The Spotlight Swings to Suez," *National Geographic*, CI (1952), 105-115.

Norse, J.E., "The Suez Canal," *Journal of the Franklin Institute*, XCI (1871), 238-242.

"The Opening of the Suez Canal as communicated to Bullion Bales, Esq, of Manchester by his friend Mr. Scamper," *Blackwood's Magazine*, CVII (1870), 85-104, 179-197, 356-375.

see Penfield listing under Panama Canal, p.

Rockwell, Charles H., "The Suez Canal," *Journal of the Franklin Institute*, LXXXV (1868), 236-242, 319-325, 377-382, "Supplement", LXXXVI (1868), 28-30.

SUEZ CANAL (cont.)

    Schoenfield, Hugh J., *The Suez Canal in Peace and War, 1869–1969* (Coral Gables, Fla: Miami Univ. Press, 1952, 1969).

    ————————————, *The Suez Canal in World Affairs* (New York: Philosophical Library, 1953).

    Taboulet, Georges, "Aux Origines du Canal de Sueq: Le Conflit Entre deLesseps et les Saint-Simoniens," *Revue Historique* [French], CCXL (1968), 89–114, 361–392.

    Three New Routes to India," "Art VIII–1. *Observations on the Euphrates Line of Communication with India* By COLONEL CHESNEY. . . 2. *The Dead Sea a New Route to India* By CAPTAIN WILLIAM ALLEN 3. *Canalization de l'Isthme de Suez* Expose DE FERD. DE LESSEPS," *North American Review*, LXXXIII (1856), 133–167.

    "The Widening of the Suez Canal," *Scientific American*, XCIX (July 25, 1908), 60–62.

    Wilson, Arnold T., *The Suez Canal, Past, Present, Future* (Oxford: Oxford Univ. Press, 1933, reprint, Arno, 1977).

    Williams, Maynard O., "The Suez Canal: Short Cut to Empires," *National Geographic*, LXVIII (1936), 611–632.

    Wright, Elizabeth Washburn, "The Canal," *Scribners*, XXXVIII (1903), 692–695.

BRITISH CANALS, selected titles

Bibliographical Aids

Baldwin, Mark, *Canal Books: A Guide to the Literature of the Waterways* (London: M & M Baldwin, 1984).

——————, "A Bibliography of British Canals, 1623-1950," in Baldwin and Burton, 130-191

——————, "Charles Hadfield: A Biographical Introduction," in Baldwin and Burton, 4-8.

Chaloner, W.H. and R. C. Richardson (comps.), *Bibliography of British Economic and Social History* (Manchester: Manchester Univ. Press, 1984).

Hadfield, Charles, "Sources for the History of British Canals," *Journal of Transport History*, II (1955), 80-89.

Skempton, Alex Westley, *British Civil Engineering, 1640-1840: A Bibliography of Contemporary Printed Reports, Plans and Books* (London: Mansell, 1986).

Other Works
British General

Baldwin, Mark and Anthony Burton (eds.), *Canals: A New Look, Studies in honour of Charles Hadfield* (Shopwyke Hall, Chichester, Sussex: Phillimore & Co., Ltd., 1984).

Barker, T.C., "The Beginnings of the Canal Age in the British Isles, (the Newry Canal)" in L. S. Presnell (ed.), *Studies in the Industrial Revolution Presented to T.S. Ashton* (1960), 1-22.

Burton, Anthony, *The Canal Builders*, Rev. Ed. (No. Pomfret, Vt: David & Charles, 1972, 1981).

Elton, Julia, *Rivers and Canals: The Science and Engineering of Waterways* (London: B. Weinreb, 1984).

Fairelough, K.R., "The Waltham Pound Lock," *History of Technology* IV (1979), 31-44.

Farrington, F.H. (ed.), "Richard Castle's 'Essay on Artificial Navigation' 1730," *Transport History*, V (1972), 67-89, 155-167.

Gladwin, D.D. & J.M., *The Canals of the Welsh Valleys and their Tramroads (CAMLASAU'R CYMOEDD A'U DRAMFFYRDD)* (n.p., The Oakwood Press, 1974).

BRITISH CANALS (CONT.)

Other British (cont.)

Hadfield, Charles, *British Canals: An Illustrated History* (Newton Abbot: David & Charles, 1974)

——————————, *The Canal Age* (New York: Frederick A. Praeger, 1969).

——————————, "Canals: Inland Waterways in the British Isles," in Singer, E.J. Holmyald, A.R. Hall and Trevor T. Williams, *History of Technlogy*, 5 vols, (Oxford Univ. Press, 1955-1958), IV; Ch 18, part II.

Phillips, J., *A General History of Inland Navigation, Foreign and Domestic* (5th ed., 1805), reprinted as *Phillips Inland Navigation* (Newton Abbot: David & Charles Reprints, 1970; N.Y: Augustus M. Kelley, 1970).

Pollins, Harold, "The Swansea Canal," *Journal of Transport History*, I (1953), 135-154.

Porteus, John D., *Canal Ports: the Urban Achievement of the Canal Age* (N.Y: Academic, 1977).

Pratt, Edwin A., *A History of Inland Transport and Communication* (1912, reprinted Newton Abbot: David & Charles, 1970).

Priestley, Joseph, *Historical Account of the Navigable Rivers, Canals and Railways Throughout Great Britain*, 2d edition (London: Frank Cass and Co., Ltd., 1831, 1967, reprint, Augustus M. Kelley, 1968).

Raven, Kate (ed.) and Jon Raven (compiler), *Canal Songs* (Tettenhall, Wolverhampton, Staffordshire: Broadside Records, 1974).

Ward, J.R., *The Finance of Canal Building in Eighteenth-Century England* (Oxford: Oxford Univ. Press, 1974)

BRITISH CANALS (Cont.)

England

Hodgson, P., "Exploring England's Canals," *National Geographic*, CXLVI (July 1974), 76-111.

Howarth, R., *Underground Canals: Wonders of the Industrial Revolution* (Ormskirk: Hesketh, 1986).

MacElwee, Roy S., "The Manchester Ship Canal," *National Waterways* VI (May 1929), 25-34, 60.

Rolt, L.T.C., *The Inland Waterways of England* (London: George Allen and Unwin, 1950).

Skempton, A.W., "The Engineers of the English River Navigation, 1620-1760," *Newcomen Society Transactions*, (English), XXIX (1953-1954 and 1954-1955), 25-54, Plates Nos. VIII, IX.

Smith, George, "Our Canal Population," *Fortnightly Review*, XXIII (1875), 233-242.

Squires, Roger W., *The New Navvies: A History of the Modern Waterways Restoration Movement* (Shopwyke, Chichester, Sussex: Phillimore & Co., Ltd., 1983).

Stephens, W.B., "The Exeter Lighter Canal, *Journal of Transport History*, III (May 1957), 1-11.

Thurston, E. Temple, *The "Flower of Gloster"* (London: Williams and Norgate, 1911; reprint, Augustus M. Kelley, 1968; reprint A. Sutton, 1984).

Willan, T.S., *River Navigation in Englsnd, 1600-1750* (Oxford: Oxford Univ. Press, 1936, reprint, Augustus M. Kelley, 1965).

FICTION

Cornish, Margaret, *Still Waters: Mystery Tales of the Canals* (London: R. Hale, 1982).

Juvenile

Smith, Peter L., *Canals are Great* (Wakefield, West Yorkshire: Author, 1977).

BRITISH CANALS (cont.)

Scotland

Cameron, Alexander Durand, *The Caledonian Canal*, 2d edition (Lavenham, Suffolk: Melven Press, 1983).

Lindsay, Jean, "The Aberdeenshire Canal, 1805-54," *Journal of Transport History*, VI (1967), 150-165.

──────────────, *The Canals of Scotland* (Newton Abbot: David and Charles, 1968).

Pratt, Edwin A., *Scottish Canals and Waterways, Comprising State Canals, Railway-owned Canals and Present-day Ship Canal Schemes* (London: Selwyn and Blount, ltd, 1922).

Thomson, G., "James Watt amd the Monkland Canal," *Scottish Historical Review*, XXIX (1950), 121-133.

Weber, D., "Caledonian Cruise," *Geographic Magazine*, LII (1980), 710-712.

BRITISH CANALS (Cont.)

Ireland

Delany, Ruth, *The Grand Canal of Ireland* (Newton Abbot: David and Charles, 1966, 1973).

Delany, Vincent T.H. and D.R., *The Canals of the South of Ireland* (Newton Abbot: David, 1972).

Malet, Hugh, *Voyage in a Bowler Hat* (London: Hutchinson & Co., 1960).

O'Hanlon, Herbert, "The Canals of Ireland," *AC*, No. 7, Nov 1973), 7;

McCarton, J., "Irish Sketches: Grand Canal," *New Yorker*, XLVII (Nov 20, 1971), 190+.

McCutcheon, W.A., *The Canals of the North of Ireland* (Dawlish, England: David and Charles, 1965; reprint, 1968).
————————, "Inland Navigation in the North of Ireland," *Technology and Culture*, VI (1965), 596–620.
————————, "The Lagan Navigation," *Irish Geography*, IV(4), 244–255.
————————, "The Newry Navigation: the Earliest Inland Canal in the British Isles," *Geographical Journal*, CXXIX (1963), 466–484.

Redford, Polly, "Cows on the Quay," *Atlantic*, CCXV (March 1965), 178–183 [Irish Canals].

Unpublished

McCutcheon, W.A.,"The Development and Subsequent Decline of the Chief Inland Waterways and Standard Guage Railways in the North of Ireland," PhD, Queen's University, Belfast, 1962.

EUROPEAN (unclassified)

Bibliographical Aid

Aldcroft, Derek H. and Richard Rodger (comps.), *Bibliography of European Economic and Social History* (Manchester: Manchester Univ. Press, 1984).

Baldwin, Mark, "A Bibliography of European Cruising, 1833-1939," in Baldwin and Burton, 85-92.

Other works

Boyer, David S., "Paris to Antwerp with the Water Gypsies," *National Geographic*, CVIII (1955), 530-559.

Calvert, Roger, *Inland Waterways of Europe* (London: Flare Books, [c. 1963] 1975).

Canal d'Entrevoches: Creer une voie navigable de la mer du Nord a la Mediterranee au XVIIe siecle [Rhone to Rhine], various authors, French and German texts (Stuttgart: Verlag Konrad Wittwer, 1987).

Cermakian, Jean, *The Moselle River and Canal from the Roman Empire to the European Economic Community* (Toronto and Buffalo: University of Toronto Press, 1975.

Chater, Melville, "Through the Back Doors of Belgium: Artist and Author Paddle for Three Weeks Along 200 Miles of Low-Country Canals in a Canadian Canoe," *National Geographic*, XLVII (May 1925), 499-540.

Chelminski, Rudolph, "The not-so-blue Danube: A storied link between Europes old and new," *Smithsonian*, XXI (July 1990), 32-42.

Farson, [James Scott] Negley, *Sailing Across Europe* (New York: Century, 1926), later described in his autobiography, *The Way of a Transgressor* (London: Gollancz, 1935).

"Floating Across the Continent," with illustrations by Pierre Boulet, pps 44-54 and Mary Leatherbee, "Fresh Bread, A Secret Agent and a Locktender Cactus," pps 63-64A in *Life* LXII (June 23, 1967.

Hammond, W.E., "The Development of the Marne-Rhine Canal and the Zollverein," *French Historical Studies*, IV (1964), .

Hlavacka, Milan "Economic and Technical Aspecs of the Elbe and Moldau Navigation in the Period of Industrial Revolution," *Hospodarske Dejiny/Economic History*, 27-47 (Prague: Institute of History of the Czechoslovak Academy of Sciences, 1990) also in German, 7-25.

EUROPEAN (unclassified) (cont.)

Hoffmann, Hanns Hubert, *Kaiser Karls Kanalbau: "Wie Kunig Carl der Grosse unterstunde de Donat und der Rhein Jusamenzugraben"* (Sigmaringen und Munchen: Jan Thorbecke Verlag, 1969).

Johnson, Irving and Electra, "Inside Europe Aboard *Yankee,*" *National Geographic,* CXXV (1964), 159-195.
——————————————, *Yankee Sails Across Europe* (New York: Norton, 1962).

Kiss, George, "TVA on the Danube," *Geographical Review,* XXXVII (1947), 274-302.

Klein, L., "Notes on the Internal Improvements of the Continent of Europe," *Journal of the Franklin Institute,* XXXIII (1842), 1-10; 73-79; 145-149; 217-225 [includes canals and railroads].

Koroknai, Akos, "Opening of the Iron Gate in 1896," *Danubian Historical Studies* [Hungary[, I (1), (1987), 39-47 [Iron Gate Canal].

Lawson, Lyle (ed.), *Waterways of Europe* [Insite Guides] (n.p: APA Publications Ltd., 1989)

McKnight, Hugh, *The Guinness Guide to Waterways of Western Europe* (Enfield, Middlesex: Guiness Superlatives, 1978).

Michel, Aloys A., "The Canalization of the Moselle and West European Integration," *Geographical Review,* LII (1962), 475-491.

Martin, J.E., "Some Effects of the Canalization of the Moselle," *Geography,* LIX (1974), 298-308.

Minshall, Merlin, "By Sail Across Europe," *National Geographic,* LXXI (1937), 533-567.

Pilkington, Roger, *Small Boat to Alsace* (New York: St. Martins, 1962).
——————————————, *Waterways in Europe: A Guide to Inland Cruising* (London: J. Murray, 1972).

Teichorn, Alice and Penelope Ratcliffe, "British Interests in Danube Navigation after 1918," *Business History,* XXVII (1985), 283-300.

Vine, P.A.L., "British Pleasure Boating on the Continent (1851-1939)," in Baldwin and Burton, 62-84.

FRANCE

Baudin, Pierre, "The Internal Navigation of france," *Contemporary Review*, LXXXIII (1893), 797-819.

Bellet, Daniel, "The Inferiority of Canals as a Means of Foreign Transport," *Engineering News*, LIV (Oct 19, 1905), 419.

Benest, E.E. (author and compiler), *Inland Waterways of France* (St. Ives: Imray, Laure, Norie and Wilson, 1965, 1971).

Bergasse, Jean-Denis, *Pierre-Paul Riquet et le Canal du Midi dans les Artes et la Litterature*, 4 vols (copyright, Jean-Denis Bergasse, 1982).

Bravard, Jean-Paul, "Mighty Rivers: Rhone Power, From the High Alps to the High Dams," *Geographical Magazine* LIX (1987), 537-542.

Un Canal. . . .Des Canaux (n.p., Caisse Nationale des Monuments Historiques et des Sites Ministere de la Culture, Picard editeur, 1986) [exposition catalogue, superb illustrations, maps, chronologies, bibliography, etc.].

Chater, Melville, "Across the Midi in a Canoe," *National Geographic*, LII (1927), 127-167.

───────────, "Through the Back Doors of France: A Seven Weeks' Voyage in A Canadian Canoe from St. Malo, Through Brittany and the Chateau Country, to Paris," *National Geographic*, XLIV (July 1923), 1-51.

───────────, *Two Canal Gypsies; their eight hundred mile Canal Voyage through Belgium, Britany, Touraine, Gascony and Languedoc* (New York: Brewer, Warren and Putnam, 1932).

Descombes, Rene, *Canaux et Batellerie en Alsace: histoire et anecdotes* (Strasbourg Illkirch: Le Verger, 1988) [illustrated].

"A Giant Among Tunnels Where the Marseilles Canal Goes Through the Mountain," *Scientific American*, CXV (Nov 25, 1916), 478-479.

Gardner, Herb, "They Shipped Her Across and Cruised the French Canals," *Cruising World*, X (Feb 1984), 60-65.

Ghilini, Hector, "Two-Seas Canal," *Living Age*, CCCLVII (1939), 178-181.

Gillmor, C. Stewart, *Coulomb and the Evolution of Physics and Engineering in the Eighteenth-Century France* (Princeton: Princeton Univ. Press, 1971).

Guillerme, Andre E., *The Age of Water: The Urban Environment in the North of France* (College Station: Texas A & M Univ. Press, 1988).

FRANCE (cont.)

Geiger, Reed, "Planning the French Canals: The 'Becquey Plan' of 1820-1822." *Journal of Economic History*, XLIV (1984), 329-339.

Grosskreutz, Helmut, *Privatkapital und Kanallan in Frankreich: 1814-1848 e. Fallstudie jur Role d. Banken in d. franz. Industrializierung* [in German] (Berlin: Duncker und Humbert, 1977).

"Inland Waterways and Waterway Projects in France," *Engineering News*, LXXVI (Nov 16, 1911), 609-610.

"The Largest Tunnel in the World [Marseilles-Rhone Canal]," *Engineering News*, LXXIV (Aug 26, 1915), 386-387.

Maistre, Andre, *Le Canal des deux mers, canal royal du Languedoc, 1660-1810* (Toulouse: Edouard Privat, 1969).

Merger, Michele, "Transport History in France," *Journal of Transport History*, 3rd series, VIII (1987), 179-201, pps 183-189, "Waterways".

Morgan, Roger, "Canals of History," *The Geographical Magazine*, LIX (1987), 360, 362-3.

Pilkington, Roger, "Pierre-Paul Ragnet and the Canal du Midi," *History Today*, XXIII (1973), 170-176.
----------------, *Small Boat Through Southern France* (London: Macmillan, 1965).

Pinsseau, Hubert, *Histoire de la Construction de l'Administration et de l'Exokiutation du Canal d'Orleans de 1676 a 1954* (Paris: Clavreuil, 1963).

Pollard, Sidney and C. Holmes (eds.), *Documents of European Economic History*, Volume One, *The Process of Industrialization, 1750-1870* (New York: St. Martins Press, 1968); Docs. #13/7 and #13/8.

Rodondi, Pietro, "Along the Water: The Genius and the Theory, D'Alembert, Condorcet and Bossut and the Picardy Canal Controversy," *History and Technology*, II (1985), 77-110.

Rolt, L.T.C., *From Sea to Sea: The Canal du Midi* (Athens. Ohio: Ohio University Press, 1973).

Roquette-Buisson, Odile de., *Le Canal du Midi* (Marsailles: Editions Rivages, 1983), [illustrated].
-----------------------------, *The Canal du Midi*, translated by Julian Guest, (London: Thames and Hudson, 1983).

FRANCE (cont.)

    Smiles, Samuel, "The Grand Canal of Languedoc and Its Constructor, Pierre-Paul Riquet de Bonrepos," reprinted from *Lives of the Engineers* (London, John Murray, 1874), I, 301-312 in Thomas Parke Hughes (ed.), *The Devolopment of Western Technology Since 1500* (New York: Macmillan Co., 1964) 39-50.

    Smith, Cecil O., Jr., "The Longest Run: Public Engineers and Planning in France," *American Historical Review*, XCV (1990), 657-692.

    Sutton, Keith, "A French Agricultural Canal-- The Canal de las Saudre and the Nineteenth-Century Improvement of the Sologne," *Agricultural History Review*, XXI (1) (1973), 51-56.

    Tunnel for Marseilles Canal, Largest in World," *Engineering News*, LXXVI (Nov 30, 1916), 1012-1014.

Unpublished

    Allen, Turner W., "The Highway and Canal System in Eighteenth Century France," PhD, University of Kentucky, 1958.

    Hannaway, John J., "The Canal of Burgundy, 1720-1854 A Study in Mixed Enterprise," PhD, Johns Hopkins, 1971.

    Gillmor, Charles Stewart, "Charles Augustin Coulomb: Physics and Engineering in Eighteenth Century France," Phd, Princeton Univ, 1968.

SPAIN

    Perez Arrion, Guillermo, *El Canal Imperial y la Navigation hasta 1812* (Zaragoza: Institucion Fernando el Catolico, 1975?) [Spanish].

    "The Cordova-Seville Canal," *American Review of Reviews*, XL (1919), 217-218.

BELGIUM

  MacElree, Roy S., "The Port of Antwerp Extension, The New Bassim Canal and Kruisschans Lock," *National Waterways*, VII (July 1929), 25-30.

  —————————, "The Ship Canal Ports of Belgium," *National Waterways*, VI (June 1929), 16-20; VII (August 1929), 24-32; VII (Sept 1929), 19-25, 62; VII (October 1929), 111-121.

NETHERLANDS

  Elder, Mimi, "Cruising Holland's Waters," *Gourmet*, XLIV (May 1984), 6-40, 174-187.

  Fochema Andreae, S.J., "The Canal Communications of Central Holland," *Journal of Transport History*, IV (1960), 174-179.

  "Great Ship Canal of the Netherlands," *Journal of the Franklin Institute*, VII (1829), 141-142.

  "The Helder or great North Holland Canal," *Journal of the Franklin Institute*, XLIV (1847), 10-13.

  Liley, J., *Barge Country: An Exploration of the Inland Waters & Great Rivers of the Netherlands* (London: Stanford Maritime, 1980).

  MacElwee, Roy S., "The Great Ymuiden Lock of the North Sea Canal," *National Waterways*, X (May 1931), 37-44, 76-78.

  —————————, "The North Sea Canal to Amsterdam," *National Waterways*, VI (April 1929), 13-21, 58-60.

  "A New Link in the Chain Binding Amsterdam to the Rhine Cities and their Hinterlands: The Recently Completed Amsterdam-Rhine Canal to be Opened by Queen Juliana on May 21," *Illustrated London News* CCXX (May 17, 1953), 836-838 [illustrations of engineering]; continued, "Aspects of Life on the New Amsterdam-Rhine Canal and the Queen Mary of the Dutch Waterways: The Human Side of a Great Engineering Achievement in the Netherlands," CCXX (June 7, 1952), 966-967.

  "Notice of a Paper on the Helder or Great Northern canal read, , ,by Mr. G.B. Jackson. . ." *Journal of the Franklin Institute*, XLIII (1847), 371-373.

  Vries, J. de, *Barges and Capitalism: Passenger Transportation in the Dutch Economy, 1632-1839* (Utrecht: HES Publishers, 1981).

GERMANY

Burkle, Fritz, *Karl August Friedrich von Duttenhofen (1758-1836) Pionier des Wasserbaus im Wurttemberg* (Stuttgart: Klett-Cotta, 1988).

Burtenshaw, D., "Fifty Years of Neckar Navigation," *Geography*, LVIII (1973), 146-150.

"Construction of the Kiel Canal Locks," *National Waterways*, X (April 1931), 14-19.

Eltzbacher, O., "The Lesson of the German Waterways," *Contemporary Review*, LXXXVI (1904), 778-797.

"Enlarging the Kaiser Wilhelm Canal, Kiel, Germanmy," *Engineering News*, LXVIII (Oct 10, 1912), 689.

Freeman, Lewis B., "Through the Kiel Canal in the Hercules," *Living Age*, CCCI (1919), 679-688.

Gunston, David, "Kiel Canal: Busiest Ship Canal in the World," *Sea Frontiers*, XXXII (Jan-Feb 1986), 46-53.

Lotz, Walther, "The Present Significance of German Inland Waterways," *Annals of the American Academy of Political and Social Science*, CCCI (1908), 246-262.

MacElwee, Roy S., "The Kiel Canal," *National Waterways*, V (Dec 1928), 11-18.

Markmann, Fritz, *Die Deutschen Flusse and Kanale. . .* (Leipzig: W Goldmann, 1942). [in German]

Pollard, Sidney and C. Holmes (eds.), *Documents of European Economic History*, Volume Two, *Industrial Power and National Rivalry, 1870-1914* (New York: St. Martins Press, 1972); Doc #2/27.

"The Teltow Canal," *Scientific American* XCV (Oct 13, 1906), 266-267, 268-269.

Toudouze, Georges G., "En Allemagne, Les Grands Travaux de Navigation Interieure," [in French] *Revue des Deux Mondes*, 7th period, volume XI (1922), 848-872.

Wheeler, W.H., "The Baltic Canal and How It Came to be Made," *Littell's Living Age*, CCIX (1896), 131-138.

ITALY

Cagli, Enrico Coen, "Inland Navigation in Italy," *National Waterways*, X (May 1931), 14-18; X (June 1931), 25-32; XI (July 1831), 14-17.

Elerlein, Harold D., "The Old Front Door to Venice [Brenta Canal]," *Travel*, LX (Dec 1932), 30-34, 52-53.

I Canali Navigabili: Costruzione e Gestione (Milano, Giuffre Editore, 1968).

MacElwee, Roy S., "The Victor Emanuel III ship Canal at Venice," *National Waterways*, VII (Nov 1929), 23-31.

SWEDEN

Adams, Phoebe-Lou, "Slow Boat on the Gota Canal," *Atlantic*, CCXVI (November 1965), 82-86.

"Dam Supported by Bascule Bridge Closes Canal Lock," *Engineering News-Recond*, LXXXIII (July 17, 1919), 116-117.

DeMare, Eric, *Swedish Cross Cut: A Book on the Gota Canal* (Malmo, Allhems Forlag, 1964).

Floyd, D.R., "The Gota Canal: Survival Problems of a Nineteenth Century Waterway in the Twentieth Century World," *Histoical Geography*, XV(1 and 2), 1-19.

Hansen, F.V., *Canals and Waterways of Sweden* (Stockholm: P.S. Norstedt, 1915).

Hendry, Peter, "Sweden's Gota Canal," *Travel*, CV (April 1956), 27-29.

Pilkington, Roger, *Small Boat Through Sweden* (New York: St. Martin's Press, 1961).

"Ship Canal from Swedish Lake to Ocean," *Engineering News*, LXXVII (February 22, 1917), 297-298.

GREECE

Hathcock, Richard, "The Corinth Canal," *Sea Frontiers*, XXX (Nov-Dec 1984), 331-334.

MacElwee, Roy S., "The Corinth Ship Canal," *National Waterways* VIII (Feb 1930), 23-27.

Richardson, Gardner, "The Auction of the Corinth Canal," *The Independent*, LXIV (1908), 857-860.

White, Horace, "The Corinth Canal," *Nation*, LIII No. 1357 (July 2, 1891), 8-9.

ROUMANIA

"Roumania Pins Hopes on New Canal," *AC*, No. 52 (February 1985), 5.

Sharman, Tim, "Canal on the Danube Delta," *Geographic Magazine*, LV (1983), 317-321.

Spubler, Nicolas, "The Danube-Black Sea Canal and Russian Control Over the Danube," *Economic Geography*, XXX (1954), 236-245.

FINLAND

Anckar, Dag, "Finnish Foreign Policy Debate: The Saimaa Canal Case," *Cooperation and Conflict* [Norway], V(4), (1970), 201-223.

Helm, Ronald A., "Finland Regains an Outlet to the Sea: The Saimaa Canal," *Geographical Review*, LVIII (1968), 167-194.

POLAND

Squires, Roger W., "Hidden Waterway Wonders [Elblaski Canal]," *AC*, Bulletin No. 62 (Aug. 1987), 8-9; see also photograph of Inclined Plane, *AC*, Bulletin No. 53 (May 1985), 2.

RUSSIA

Axelbank, A., "Letter from Ashkhabad," *Far Eastern Economic Review*, XCIX (Feb 3, 1978), 54.

Bakshy, Alexander, "Russia's New outlet to the Sea," *Current History*, XLI (1935), 562-566.

Blackwell, William L., *The Beginnings of Russian Industrialization, 1800-1860*, 2 vols (Princeton, N.J:, Princeton Univ. Press); "Canals and Highways," I, 264-270.

RUSSIA (cont.)

"Canals and other Means of Transport in Russia," *Geographical Journal*, XLIII (1914), 577-578; excerpted from R. Hennig, in *Deutsche Geogr. Blatter*, XXXVI, pts 3-4.

Cross, Anthony, "A Russian Engineer in Eighteenth-Century Britain: the Journal of N. I. Korsakov- 1776-7," *Slavonic and East European Review*, LV(1), (1977), 1-20.

Gorky, Maxim, L. Auerback and S. G. Firin, *Belomor: An Account of the Construction of the New Canal Between the White Sea and the Baltic Sea* (N.Y: Harrison Smith and Robert Haas, 1935, reprint, Hyperion, 1977).

Haywood, Richard M., *The Beginnings of Railroad Development in Russia in the Reign of Nicholas I, 1835-1842* (Durham: North Carolina, Duke Univ. Press, 1969), pps. 3-21, 31-34.
————————————, "The Development of Steamboats on the Volga River and its Tributaries," *Research in Economic History*, V (1981), 127-192.

Istomina, E.G., "The Volga Water Route in the Second Half of the 18th and the Beginning of the 19th Century," *Soviet Geography*, XXVIII (May 1987), 330-350.

Jones, Robert E., "Getting the Goods to St. Petersburg: Water Transport from the Interior 1703-1811," *Slavic Review*, XLIII (1984), 411-433.

Kerner, Robert J., *The Urge to the Sea: The Course of Russian History* (Berkley and Los Angeles: Univ. of California Press, 1942).

Latrell, C., "Gorky as Apologist: The White Sea Canal Project," *Yale Theater*, VII (Winter 1976), 88-94.

Matley, Ian M. "The White Sea-Baltic Canal: a route to the Russian north," *Journal of Transport History*, 3rd series, XI (March 1990), 29-39.

"Russian Inland Navigation," *Journal of the Franklin Institute*,

PHILIPPINES

Mallari, Francisco, "Nineteenth Century Spanish Bureaucracy: A Case Study," *Philippine Studies*, XXXI(3), (1983), 382-396.

INDIA

Brueckmann, Dee and Korte, "India's Backwater Highways," *Oceans*, XX (Jan-Feb 1987), 24-29.

Harnetty, Peter," 'India's Mississippi': The River Godavari Navigation Scheme," *Journal of Indian History*, XLIII (1965), 699-732.

Moncrieff, G.K. Scott, "On An Indian Canal," *Blackwood's*, CLXXXIII (1908), 782-793; CLXXXIV (1908), 39-47, 191-199.

Trout, William E., "A Canal Wallah in India," *AC*, Part I, No. 49 (May 1984), 6-7; Part 2, No. 50 (August 1984), 6-7, 9.

JAPAN

Tanare, Saguro, "The Lake Biwa-Kioto Canal, Japan," *Scientific American*, LXXV (Nov 7, 1896), 342, 345-6.

Trout, William E., "A True Account of the Adventures of an American on Japan's Biwako Canal in the Year of the Monkey," *The Towpath Post* [Canal Society of New Jersey], (Summer 1974), 1-12.

THAILAND

Duncan, R.D., "Proposed: A Kra Canal," *United States Naval Institute Proceedings*, XC (1964), 48-55.

Libby, N. F., "Thailand's Kra Canal: Site for the World's First Nuclear Industrial Zone," *Orbis*, XIX (1), (1975), 200-208.

Ronan, William J., "The Kra Canal: A Suez for Japan," *Pacific Affairs*, IX (1936), 406-415.

Shigeharu. Tanabe, "Historical Geography of the Canal System in the Chao Phraya River Delta," *Journal of the Siam Society*, LXV (2), (1977), 23-72,

Trout, William E., "Canals in Thailand," *AC*, No. 42 (August 1982), 6.

MADAGASCAR

Hance, William A., "Transportation in Madagascar," *Geographical Review*, XLVIII (1958), 45–68 [particularly 55–56].

AUSTRALIA

Fenner, Charles, "The Murray River Basin," *Geographical Review*, XXIV (1934), 79–91.

VIETNAM

Daniel, Mann, Johnson & Mendenhall International, *Final Report of an Engineering Study of the Primary Canal System of South Vietnam, with a Preliminary Reconnaissance of the Saigon Ship Canal* (Saigon: 1960).

COLOMBIA

Kelley, F.M., "On the Junction of the Atlantic and Pacific Oceans and the Practicability of a Ship Canal, withour Locks, by the Valley of the Atrato," *Journal of the Franklin Institute*, LXII (1856), 83–89.

Nichols, Theodore, "Cartegena and the Dique: A Problem in Transportation," *Journal of Transportation History*, II (1955–56), 22–34.

Serrell, Edward W., "Inter-oceanic Ship Canal via the Atrato and Truando Rivers," *Journal of the Franklin Institute*, LX (1855), 289–291.

Trautwine, John C., "Remarks Concerning Surveys of the Atrato River," *Journal of the Franklin Institute*, LXXIV (1862), 27–31.
——————————, "Rough Notes of an Expedition for an Inter-oceanic Canal Route by way of the Rivers Atrato and San Juan in New Grenada, South America," *Journal of the Franklin Institute*, LVII (1854), 145–154; 217–231; 289–299; 361–373; LVIII (1854), 1–12; 73–84; 145–155; 216–226; 289–299 [with maps and numerous unusual illustrations]

MEXICO

Hough, Walter, "The Venice of Mexico," *National Geographic*, XXX (1916), 69–88.

SOUTH AMERICA (General)

Montero, Homero Martinez, "Towards a South American Canal," *Americas*, XVIII (Sept 1966), 23–27.

CHINA

Bates, M.S., "Problems of Rivers and Canals under Han Wu Ti, 140-187 B.C.," *American Oriental Society Journal*, LV (1935), 303-306.

Ch'ao-Ting Chi, *Key Economic Areas in Chinese History as Revealed in the Development of Public Works for Water-Control* [1936] (reprint, N.Y: Augustus M. Kelley, 1970).

Deering, Mabel Craft, "Ho for Soochow Ho," *National Geographic*, LI (June 1927), 623-649.

Gander, Le P. Domin, S.J., *Le Canal Imperial Etude Historique et Descriptive, Varietes Sinologiques*, No. 4, 1894 [Kraus Reprint, 1975], in French.

The Grand Canal of China published by South China Morning Post and New China News, c. 1985; briefly reviewed in *Far Eastern Economic Review*, CXXVII(9), (March 7, 1985), 49.

Harrington, Lyn, *The Grand Canal of China* (Chicago: Rand McNally, 1967).
————————, "The Grand Canal of China: Longest of all man-made waterways has been used for nearly 2,500 years," *Natural History*, LXXV (Aug/Sept 1966), 16-21.

King, F. H., "The Wonderful Canals of China," *National Geographic*, XXIII (1912), 931-958.

Leonard, J.K., "'Controlling from afar' open communications and the Tao-Kuang Emperor's control of grand canal-grain transport management," *Modern Asian Studies*, XXII (Oct 1988), 665-699.

Needham, Joseph, "China and the Invention of the Pound Lock," *Transactions of the Newcomen Society* [English], XXXVI (1963-1964), 85-107.
————————, *Science and Civilization in China*, vol IV *Physics and Physical Technology*, Part III "Civil Engineering and Nautics," 269-365, plates Nos. 899-904, 917-925.

Pin, Liao (ed.), *The Grand Canal: An Oddessy* (Foreign Language Press, 1987)

Price, Willard, "Grand Canal Panarama," *National Geographic*, LXXI (Apr. 1937), 487-514.

CHINA(cont.)

    Squires, Roger W., "A Visit to China," *AC*, No. 71 (November 1989), 8-9.

    Trout, William E., III, "The Emperor's Lock Model," *AC*, No. 40 (February 1982), 3, 4; reprinted in *The Best from American Canals*, II, 84.

Unpublished

    Huang, Ray, "The Grand Canal During the Ming Dynasty, 1368-1644," PhD, University of Michigan, 1964).

SUDAN

    Collins, Robert O., *The Waters of the Nile: Hydropolitics and the Jonglei Canal 1900-1988* (New York: Oxford Univ. Press, 1990)

    Eshman, Robert, "The Jonglei Canal: A Ditch Too Big?" *Environment*, XXV (June 1983), 15-20, 32.

    Howell, P.P., "The Impact of the Jonglei Canal in the Sudan," *Geographic Journal*, CXLIX (1983), 286-300.

    Howell, Paul, Michael Lock and Stephen Cobb (eds.) *The Jonglei Canal: Impact and Opportunity* (Cambridge et al: Cambridge Univ. Press, 1988)

    Waterbury, John, *Hydropolitics of the Nile Valley* (Syracuse, N.Y., Syracuse Univ. Press, 1979).

    Wright, John, "Sudan Holds the Key," *Geographic Magazine*, LI (1978), 33-42 [proposed loop on the Nile].

ANCIENT

Adams, Richard E.W., "Ancient Maya Canals: Grids and Lattices in the Maya Jungle," *Archaeology*, XXXV (Nov/Dec 1982), 28-35.

Bonnard, Louis, *La navigation interieure de la Gaul a l'epoque gallo-romaine* (Paris: A. Picard et fils, 1913) [in French].

Brittain, Robert, *Rivers, Man and Myths: From Fish Spears to Water Mills* (Garden City, N.Y: Doubleday & Co., 1958), numerous scattered canal references.

DeCamp, L. Sprague, *The Ancient Engineers* (Garden City, N.Y: Doubleday & Company, 1963).

Eckoldt, Martin, "Navigation on Small Rivers in Central Europe in Roman and Medieval Times," *International Journal of Nautical Archaeology and Underwater Exploration,*" XIII (February 1984), 3-10.

Eubanks, J.E., "Navigation on the Tiber," *Classical Journal*, XXV (1929-1930), 683-695.

Forbes, R.J., *Studies in Ancient Technology*, 9 vols (Leiden: E.J. Brill, 1955-   ), especially vol II, "Dikes, Windmills and Sluices," 53-71, "Chronological Sy of Irrigation and Drainage Data," II, 72-77.

Gest, Alexander P., *Engineering* (New York: Cooper Square Publishers, Inc., 1963), 188-196.

"A Historical Account of the Canal Which Connected the Nile and the Red Sea in Ancient Times," *Blackwood's Magazine*, LVI (1844), 182-194.

Howarth, S., "Atrato Canal Revealed [Colombia, SA]," *Geographical Magazine*, XLIV (1971), 420-3.

Lloyd, R.G., "Aqua Virgo, Euripus and Pops Agrippae," *American Journal of Archaeology*, LXXXIII (April 1979), 193-204.

Leger, Alfred, *Travaux Publies aux Temps des Romains* (Nogent-le-Roi: Jacques Laget, 1979).

MacDonald, Brian R., "The Diolkos," *The Journal of Hellenic Studies*, CVI (1986), 191-5.

Moore, Frank G., "Three Canal Projects, Roman and Byzantine," *American Journal of Archaeology*, LIV (1950), 97-111.

ANCIENT (cont.)

Sasel, J., "Trajan's Canal at the Iron Gate," *Journal of Roman Studies*, LXIII (1973), 80–85.

Scarborough, Vernon L., "A Preclassic Maya Water System," *American Antiquity*, XLVIII (1983), 720–740.

Siemens, Alfred H. and Dennis E. Puleston, "Ridged Fields and Associated Features in Southern Campeche: New Perspectives on the Lowland Maya," *American Antiquity*, XXXVII (1972), 228–239; "Commentary on Lowland Maya," *American Antiquity*, XLI (1976), 381–384.

Smith, N.A.F., "Roman Canals," *Newcomen Society Transactions* [English], XLIX (1977–1978), 75–86.

Sneh, Amihai, Tuvia Weissbrod and Itamar Perath, "Evidence for an Ancient Egyptian Frontier Canal," *American Scientist*, LXIII (1975), 542–548.

Verdelis, Nicholas M., "How the Ancient Greeks transported Ships over the Isthmus of Corinth: Uncovering the 2550-year-old *Diolkis* of Periander," *Illustrated London News*, CCXXXI (19 Oct. 1957), 649–651.

Westermann, William Linn, "On Inland Transportation and Communication in Antiquity," *Political Science Quarterly*, XLIII (1928), 364–387 [particularly 382–385]; another version in *Classical Journal*, XXIV (April 1929), 493–497 [492–496].

MEDIEVAL AND RENAISSANCE

Eckoldt, Martin, *Schiffahrt auf kleinen Flussen Mitteleuropas in Romerzeit und Mittelalter* (Oldenburg, Hamburg, Munchen: Stalling Verlag GmbH, 1980). [German]

Fasso, Constantino A., "Birth of hydraulics during the Renaissance period," in Gunther Garbrecht (ed.), *Hydraulics and Hydraulic Research: A Historical Review* (Rotterdam/Boston: A.A. Balkema, 1987), 55-79, "Birth of Navigation Locks," pps. 71-74.

Gille, Bertrand, *Engineers of the Renaissance* (1964; Cambridge, Mass: M.I.T. Press, translated 1966).

Leighton, Albert C., *Transport and Communication in Early Medieval Europe, AD500-1100* (Newton Abbot: David & Charles, 1972), pps. 128-130.

Lieb, John W., "Leonardo Da Vinci--Natural Philosopher and Engineer," in Thomas Parke Hughes, *The Development of Western Technology since 1500* (New York: Macmillan, 1964) 27-32; reprinted from the *Journal of the Franklin Institute*, CXCI (1921), 767-806; CXCII (1921), 47-68.

Parsons, William Barclay, *Engineers and Engineering in the Renaissance* (Baltimore: The Williams and Wilkins Co., 1939). part 5, chap. 22-27 and appendix C(4).

APPENDIX A
[compiled with the aid of Michael Knies,
Collections Manager, Hugh Moore Historic Park and Museums

The following is a very incomplete and not necessarily up-to-date list of organizations concerned in some way with canal history, preservation or promotion. It is offered for what it is with the solicitation that users of this volume will send additions, corrections, up-dates, etc.

Akron, Ohio, Lock II Park
Alexandria Canal Museum, c/o Historic Alexandria
    405 Cameron St., Alexandria, VA 22314.
        restoration, museum and library in progress
Allegheny Portage Railroad National Historic Site
    Rt. 22, P.O. Box 247, Cresson, PA 16630 [814-886-8176]
        Canal museum, restored inclined plane & publications
Allen County - Fort Wayne Historical Society
    302 East Berry St., Fort Wayne, IN 46802
        Canal museum and publications.
American Canal & Transportation Center- Pennsylvania
    809 Rathton Rd., York, PA 17403 [717-843-4035]
        Historic Pennsylvania Canal, Railroad and highway books.
American Canal & Transportation Center- West Virginia
    P.O. Box 310, Shepherdstown, WV 25443
        Historic Maryland canal books.
American Canal Society
    117 Main Street, Freemansburg, PA 18017
        American Canals and other publications
Association for Great Lakes Maritime History
    P.O. Box 25, Lakeside, OH 43440
        newsletter
B.A.R.G.E.- Bi-Recreational Association for the Resoration of the
                  Great Erie
    72 Harvington Rd., Tonawanda, NY 14150
        The Canawler and Erie Canal boating.
Barron's C & O Canal
    P.O. Box 356, Snyders Landing Road, Sharpsburg, MD 21782
        [301-432-8726]
Blackwells Mills Canal House Association
    RD #1, Box 53, Sommerset, NJ 08873
        on Delaware and Raritan Canal
C&O Canal Association
    Box 366, Glen Echo, MD 20812-0366
        Along the Towpath, field trips and educational programs.
C&O Canal National Historic Park
    P.O. Box 4, Sharpsburg, MD 21782
        The Towline, restoration and boat rides.
C&O Cumberland Inc.,
    Box 1378, Cumberland, MD 21502
        canal boat replica

C.A.N.A.L. Inc.
    36 Lakeview Drive, Lincoln, RI 02866
Camillus-Erie Canal Project
    109 East Way, Camillus, NY 13031
        canal restoration work
Canadian Canal Society
    P.O. Box 1652, St. Catharines, Ontario, Canada L2R 7K1
        periodical and field trips
Canal Archives and Regional History Collection
    Lewis University, Rte. 53, Romeoville, IL 60441 [815-838-0500]
C. Howard Heister Canal Center
    c/o Berks County Parks & Recreational Department
    R.D. 5, Box 272, Sinking Spring, PA 19608
        museum
Canal Fulton Heritage Society
    P.O. Box 584, Canal Fulton, OH 44614
        Newsletter, canal boat rides, museum and canal-era house
The Canal Museum (New York)
    318 Erie Blvd. E., Syracuse, NY 13202 [315-471-0593]
        Canal Packet, museum, publications, field trips and
        restoration
Canal Society of Indiana
    302 East Berry Street, Fort Wayne, IN 46802
        Indiana Waterways and field trips.
Canal Society of New Jersey
    Maccullough Hall, P.O. Box 737, Morristown, NJ 07960
        On the Level, canal museum and field trips.
Canal Society of New York State
    311 Montgomery Street, Syracuse, NY 13202 [315-428-1862]
        Bottoming Out, canal boat trips and field trips
Canal Society of Ohio
    550 Copley Rd., Akron, OH 44320
        Towpaths, newsletter, archives and field trips
Canastota Canal Town
    122 Canal Street, Canastota, NY 13032
        museum, restoration and news letter
Carroll County Wabash & Erie Canal Inc.,
    P.O. Box 255, Delphi, IN 46923
        canal park
Chenango Canal Society
    c/o Chenango Planning Board
    99 North Broad Street, Norwich, NY 13815
Chesapeake & Delaware Canal Museum
    The Old Lock Pumphouse, Chesapeake City, MD 21915
        museum
Columbia River Projects [canalized river]
    for information
        U.S. Army Corps of Engineers, Portland District
        Portland, Oregon 97208

Corps of Engineers, Editor *Engineer Update*
    DAEN-PAC, Washington, DC 20314
        canal and waterways operations, parks.
Cumberland & Oxford Canal Association
    Raymond, ME 04071
        newsletter and guidebook
Delaware and Hudson Canal Historical Society
    300 Ohioville Rd., New Paltz, NY 12561
        newsletter and museum
Delaware and Hudson Canal Historical Society Museum
    P.O. Box 23, Mohonk Road, High Falls, NY 12440 [914-255-1538]
        canal museum
Delaware & Lehigh Navigation Canal National Heritage Corridor
    for information contact
        Deirdre Gibson
        Division of Park and Resource Planning, National Park Service
        260 Custom House, 2nd & Chestnut Sts., Phila., PA 19106
        [215-597-6486]
Delaware & Raritan Canal Coalition
    1108 Princeton-Kingston Rd., Princeton, NJ 08540
        canal protection
Delaware & Raritan Canal Commission
    Prallsville Mills, P.O. Box 539, Stockton, NJ 08559-0539
        Publications, programs.
Farmington Canal Corridor Association
    c/o Mrs. Ruth Hummel and Melvin Schneidermeyer
    Box 24 Plainville, CT 06062
Fort Hynter Canal Society
    Fort Hunter, NY 12069
        newsletter
Friends of the Delaware Canal
    P.O. Box 312, Point Pleasant, PA 18950
        newletter, restoration activities.
Friends of the I&M Canal National Heritage Corridor
    P.O. Box 867, Ottawa, IL 61350
        The I&M News
Georges River Canal Association
    R.F.D. 1, Warren, ME 04864
Historic Saltsburg, Inc.,
    Saltsburg, PA 15681
        canal park, publications and slide-sound show
Hugh Moore Historical Park and Canal Museums
    P.O. Box 877, Easton, PA 18044   215-250-6700
        The Locktender, museum, restorations, research center
            and canalboat rides.
Illinois & Michigan Canal National Heritage Corridor Civic Center
                Authority
    P.O. Box 501, Willow Spring, IL 64480
        newsletter

Also out of Illinois & Michigan Corridor
>*The I&M News* produced by friends of the corridore
and
>*Corridor Courier*, prepared by George D. Berndt, Interpretive Specialist, I&M Canal NHC, (815) 740-2049 since 1986.

Illinois & Michigan Canal Museum
> 803 South State Street, Lockport, IL 60441
>> museum and newsletter

Illinois Canal Society
> 1109 Garfield Street, Lockport, IL 60441
>> publications

Landsford Canal State Park, off US 21 fifteen miles south of Rock Hill, So. Carolina.

Lebanon County Historical Society
> Lebanon, PA 19533
>> maintains Union Canal Tunnel

Leesport Lock House Foundation
> Leesport, PA 19533
>> newsletter and restoration

Lowell National Historical Park
> 169 Merrimack St, Lowell, MA 01852
>> restoration and boat rides

Manayunk Canal Committee
> P.O. Box 4644, Philadelphia, PA 19127
>> canal restoration

Miami & Erie Canal Society
> 702 Columbia St., St. Marys, OH 45885

Middlesex Canal Association
> P.O. Box 333, Billerica, MA 01821.
>> *Towpath Topics* and field trips

Monongahela River Buffs Association
> Greensboro, PA 15338
>> *The Voice of the Mon*, museum and field trips.

Neversink Valley Area Museum
> Box 263, Cuddebackville, NY 12729
>> *Towpath Chatter*, D & H canal museum and restoration

Old Freemansburg Association
> 117 Main Street, Freemansburg, PA 18017
>> *The Locktender's Log*, canal and lock restoration

Panama Canal Commission
> Office of Public Affairs
> APO Miami, FL 34011
>> *The Panama Canal Spillway* and canal operation.

Panama Canal Commission Library
> 2000 L St., NW, Ste. 550, Washington, DC 20036 [202-634-6441]

Pennsylvania Canal Society
> P.O. Box 877, Easton, PA 18044
>> *Canal Currents*, publications, collections, field trips.

Pennsylvania State Archives
    Box 1026, Third & Forster Sts., Harrisburg, PA 17108 717-787-2701
Piqua Historical Museum
    Schmidlapp Building, 507 Main Street
    Piqua, Ohio 45356
Plainville Historic Center
    Farmington Canal Room
    P.O. Box 464, Plainville, CT 06062
        museum and publications
Portage Canal Society
    529 W. Cook St., Portage, WI 53901 [608-742-2889]
        annual letter, canal restoration and publication.
Rideau Canal - Superintendent
    12 Maple Avenue North, Smith Falls, Ontario, Canada K7A 1Z5
        Steam & Stone, canal operation, restoration, museum and
                                programs
Roscoe Village Foundation
    381 Hill St, Coshocton, OH 43812 [614-622-9310 and 2222]
        Roscoe Village News, restoration and canal boat rides
Sand;y & Beaver Canal Inc.,
    496 Carrollton St., Magnolia, OH 45643
        canal restoration
Schuylkill Canal Association,
    Box 3, Mont Clare, PA 19453
        Schuylkill Canal News
Scioto Valley Canal Society
    Box 502, Portsmouth, OH 45662
S.C.O.W - State Conference on Waterways
    493 Broadway, Sarasota Springs, NY 12866
St. Catharines Historical Museum
    343 Meritt Street, St Catharines, Ontario, Canada, L2T 1K7
        museum, lectures, field trips and publications
St. Lawrence Seaway Authority
    360 Albert St., Ottawa, Ontario, Canada K1R 7X7
        publications, pleasure craft guide and general information
St. Lawrence Seaway Development Corporation
    P.O. Box 520 - Seaway Circle, Massena, NY 13662
Schuylkill Canal Association
    1301 Black Rosk Road, Box G
    Oaks, PA 17456
Susquehanna Museum of Havre de Grace
    P.O. Box 253, Havre de Grace, MD 21078
        museum, restoration and publications
Trent-Severn Waterway
    P.O. Box 567, Ashburnham Drive, Peterborough, Ontario, Canada,
            K9J 6Z6
        operating canal

Virginia Canals and Navigation Society
    c/o Russ Harding, historian/archivist
    Box 80 RFD Route 4, Mineral, VA 23117 [703-894-5703]
  also   35 Towana Rd.,
    Richmond, VA 23226 [804-288-1334]

Walnutport Canal Association
    606 Washington St., Walnutport, PA 18088
        restoration and *Canal News*

Waterloo Village
    Stanhope, NJ 07874 [201-347-0900]
        canal restoration, museum of Canal Soceity of New Jersey

Whitewater Canal Historic Site, between Laurel and Brookville,
    Franklin County, Indiana--horse drawn canal boat at Matamora.

Woburn, Massachusetts, Historic Canal Park [Middlesex Canal]

SOME OVERSEAS SUGGESTIONS

Black Country Museum
    Tipton Road
    Dudley, West Midlands, England [phone 021 557 9643]
        includes canal boat ride with sound and light show in Dudley
            canal mine tunnel

The Boat Museum
    Ellesmere Port
    South Wirral, Cheshire, England [phone066 33 3411]
        includes canal boat ride

British Waterways Board
    Melbury House, Melbury Terrace
    London NW1 6JX England

Inland Waterways Association
    Dept CR89/4, 114 Regents Park Road,
    London, NW18UQ, England

Inland Waterways Association of Ireland
    Stone Cottage, Claremont Road
    Killiney Co.. Dublin, Ireland

Trollhatte Kanalverks Museum
    Trollhattan, Sweden

APPENDIX B
CANAL CRUISES-North America

NOTE: Disclaimer, the following are a highly selective, somewhat haphazard and not necessarily up to date listing. Any information that would up-date our next edition of this bibliography would be appreciated

American Cruise Lines, Haddam, Connecticut.
    east coast-inland waterway-Lake Okeechobee cruises

American Canadian Carribean Line, Inc., Warren, Rhode Island.
    L.I. Sound, Hudson R., Erie Canal, St Lawrence cruise [800-556-7450].
    Also Intercoastal Waterways and Florida Canal Cruises

Big Rideau Tentals
    Rideau Ferry, Ontario, Canada [613 825 3317]

Champlain Gateway Cruises
    P.O. Box 201
    Whitehall, New York 12887 [518-499-1600

Delta Queen Steamship Co, Cincinnati, Ohio.
    operates *Delta Queen* and *Mississippi Queen* on
    canalized Mississippi and Ohio Rivers.

Everglades Jungle Cruises, Fort Myers, Florida
    irregular Fort Myers-Lake Okeechobe cruises

Gray Line Water Sightseeing -- Seattle Waterfront Ship Canal - Lake Union
    via Hiram Chittenden Locks Passage Cruise
    500 Wall Styreet, Suite 413, Seattle, WA 98121 [206-441-1887]

Mid-Lakes Navigation Co, Skaneateles, N.Y.
    Three day cruises on Erie and Champlain Canals.
    also house boat rentals.

Muskola Lakes Navigation & Hotel Company, Muskola, Ontario, Canada
    two day steam cruises on Lake Muskola.

New Hope Barge Company
    P.O. Box 164, New Hope, PA 18938 [215-862-2842]
        mule drawn boat rides

Ontario Waterway Cruises, Inc.,
    Box 1540, Peterborough, Ontario, Canada K9J 7H7 [705-748-3666]
    Trent-Severn Waterway and Rideau Canal Cruises.

Paddleford Steam Packet Co of St. Paul
    summer, Mississippi Cruises
    winter, Ft. Myers-Lake Okeechobee-West Palm Beach cruises.

Riverboat Cruises, Troy, N.Y.
    day cruises, dinner cruises.

Summer Magic Houseboats
    Port Hope, Ontario, Canada   [416 885 9503]
        houseboat rentals on Rideau Canal

Spirit of the Fox, excursion boat out of "downtown Menasha [Wisconsin] marina".

NOTE: Canal Boat Rides, mostly mule drawn are also available at the C&O National Historical Park, at Canal Fulton, at the Hugh Moore Historical Park and Canal Museums, at the Roscoe Village Foundation and at The Whitewater Canal Historic Site. See above entries.

European Canal Cruising

GREAT BRITAIN

There are numerous boat liveries in Great Britain for self-operated canal boats. The two listed below are those used by the compiler.

Anglo-Welsh Waterway Holidays
5 Canal Basin
Market Harborough, Leicester LE16 7BJ England [ph. 0858 66910]

English Country Cruises
Wrenbury Mill near
Nantwich, Cheshire CW5 8HG, England [phone 0270 780544

Other boat liveries as well as operators of hotel boats and other cruise options can be found in the advertisements of the various travel magazines, throught British magazines like *Waterways World* or in *British Heritage* or by contacting travel agents. Boat rentals, cruises, hotel boats, etc. for the countries of the continent can be found in a similar manner.

IRELAND

Celtic Canal Cruisers, Ltd
Tullamore Co. Offaly, Ireland [phone 010 353 506 21861]

GERMANY

Compass Tours Incoming gmbh
Barbarossawall 11-23
4000 Dusseldorf 31, Germany [phone 0211/40 70 21
Weser & Elbe Rivers, Elbe Lateral Canal and Mitteland Canal

KD German Rhine Line
Suite 317, 170 Hamilton Ave, White Plains, NY 10601 [800-346-6525]
Rhine River and prospective Elbe River Cruises.

SWEDEN

Dalslands Kanal AB
Box 181, S-66200 Amal Sweden [phone 0532/14366]

Gota Kanal
Rederiaktiebolaget Gota Kanal
Box 272 S-401 24 Goteborg, Sweden [phone 031 17 76 15]

West Line
Rederi AB, Box 103, S 401 21 Goteborg, Sweden [ph. 031 40 12 20]
day trips on Trollhattan Ccanal.